Statistical Analysis of Reliability Data

OTHER STATISTICS TEXTS FROM
CHAPMAN & HALL

Further information on the complete range of Chapman & Hall statistics books is available from the publishers.

Statistical Analysis of Reliability Data

M. J. Crowder

Senior Lecturer
University of Surrey,
UK

A. C. Kimber

Lecturer
University of Surrey,
UK

R. L. Smith

Professor of Statistics
University of North Carolina,
USA

and T. J. Sweeting

Reader in Statistics
University of Surrey,
UK

CHAPMAN & HALL
London · New York · Tokyo · Melbourne · Madras

UK	Chapman & Hall, 2-6 Boundary Row, London SE1 8HN
USA	Chapman & Hall, 29 West 35th Street, New York NY10001
JAPAN	Chapman & Hall Japan, Thomson Publishing Japan, Hirakawacho Nemoto Building, 7F, 1-7-11 Hirakawa-cho, Chiyoda-ku, Tokyo 102
AUSTRALIA	Chapman & Hall Australia, Thomas Nelson Australia, 102 Dodds Street, South Melbourne, Victoria 3205
INDIA	Chapman & Hall India, R. Seshadri, 32 Second Main Road, CIT East, Madras 600 035

First edition 1991

© 1991 M. J. Crowder, A. C. Kimber, R. L. Smith and T. J. Sweeting
Typeset in 10/12 Times by Interprint Ltd., Malta
Printed and bound in Great Britain by
T J Press (Padstow) Ltd, Padstow, Cornwall

ISBN 0 412 30560 7

British Library Cataloguing in Publication Data
Statistical analysis of reliability data.
– (Statistics texts)
I. Crowder, M. J. II. Series
519.5

ISBN 0–412–30560–7

Library of Congress Cataloguing-in-Publication Data
Statistical analysis of reliability data / Martin Crowder ... [et al.].
— 1st ed.
    p. cm.
    Includes bibliographical references and indexes.
    ISBN 0–412–30560–7
    1. Reliability (Engineering) — Statistical methods. 2. Reliability (Engineering) — Data processing. I. Crowder, M. J. (Martin J.), 1943–   .
TA169.S73 1991
620'.00452 – dc20   91-25
                                                          CIP

Contents

Preface

Books on reliability tend to fall into one of two categories. On the one hand there are those that present the mathematical theory at a moderately advanced level, but without considering how to fit the mathematical models to experimental or observational data. On the other hand there are texts aimed at engineers, which do deal with statistics but which largely confine themselves to elementary methods. Somewhere in between these two extremes lie the books on survival data analysis, which are more up to date in their statistical methodology but which tend to be oriented towards medical rather than reliability applications.

We have tried in this book to combine the best features of these different themes: to cover both the probability modelling and the statistical aspects of reliability in a book aimed at engineers rather than professional statisticians, but dealing with the statistics from a contemporary viewpoint. In particular, we see no reason why reliability engineers should be afraid of exploiting the computational power that is present in even very small personal computers these days.

One can discern a number of reasons for the bias in existing texts towards elementary statistical methods. One is that, until quite recently, university degree courses in engineering contained little or no training in statistics. This is changing very rapidly, partly as a result of the increasing mathematical sophistication of engineering generally, and partly because of the more specific current interest in the design of industrial experiments. A second reason is a distrust of the use of complicated mathematics to solve reliability problems. Authors such as O'Connor (1985), on what he termed 'the numbers game', have provided eloquent demonstration of the dangers in following an automated statistical procedure without thinking about the assumptions on which it is based. However, there is another side to the statistical story: modern statistics is concerned not only with automated methods for fitting complicated models to data, but also with an ever growing array of techniques for checking up on the assumptions of a statistical model. Our aim here is to present a balanced account presenting not only the mathematical techniques of model fitting, but also the more intuitive concepts (frequently depending heavily on computer graphics) for judging the fit of a model.

Although we have tried as far as possible to make this book self-contained, the book is really aimed at someone with a background of at least one course in statistics, at about the level of Chatfield (1983). This will include the more common probability distributions, estimation, confidence intervals and hypothesis testing, and the bare bones of regression and the analysis of variance. Given that degree courses in engineering nowadays usually do include at least one course in elementary statistics, we do not believe that these are unreasonable requirements. The present text may then be thought of as providing a follow-up course aimed at one specific, and very common, area of the application of statistics in engineering. Familiarity with one of the more common statistical packages, such as Minitab or SPSS^X, will certainly help the reader appreciate the computational aspects of the book, but we do not assume any prior knowledge of reliability theory or life data analysis.

Chapters 1 and 2 present elementary material on probability models and simple statistical techniques. Although some of this is revision material, these chapters also introduce a number of concepts, such as the Kaplan–Meier estimate, which are not often found in introductory courses in statistics. The core of the book starts with Chapter 3, in which numerical maximum likelihood and model-checking techniques are described from the point of view of fitting appropriate distributions to samples of independent, identically distributed observations, but allowing censoring. In Chapter 4, these methods are extended to parametric regression models. The ideas presented here are familiar enough to applied statisticians, but they are not as yet widely used in reliability. This chapter includes several substantial examples which, apart from serving to illustrate the mathematical techniques involved, touch on many of the practical issues in application to reliability data. Chapter 5 is about another class of regression models: semiparametric models based on the proportional hazards assumption. Although by now widely used on medical data, these methods have not reached the same level of acceptance in reliability. In many respects, they may be thought of as an alternative to the parametric techniques in Chapter 4.

Chapter 6 is about Bayesian methods. At the present time there is lively debate over the place of Bayesian statistics in reliability theory. Whether Bayesian statistics will eventually supplant classical statistics, as its more vigorous proponents have been proclaiming for the past forty years, is something still to be seen, but it is certainly our view that reliability engineers should have an awareness of the Bayesian approach and this is what we have tried to provide here. Chapters 7 and 8 deal with more specialized themes: Chapter 7 on multivariate reliability distributions and Chapter 8 on particular methods appropriate for repairable systems. Finally, Chapter 9 presents an overview of models for system reliability, divided between the theory of coherent systems and a more specialized topic, not previously covered in a book of this nature, that of ioad-sharing systems.

Within this wide scope of topics it is inevitable that the mathematical level of the book varies appreciably. Chapters 1 and 2 are quite elementary but parts of the later chapters, especially Chapter 9, are considerably more advanced. To aid the reader trying to judge the level of the material, we have indicated with an asterisk (*) certain more advanced sections which can be omitted without loss of continuity.

One area we have not tried to cover is experimental design. The revolution in engineering design associated with the name of Taguchi has focused considerable attention on the proper design of industrial experiments. To do justice to this here would have taken us too far from our main theme, but we certainly do regard experimental design as an important part of reliability, especially in connection with the more advanced analytic techniques described in Chapters 4 and 5. We would encourage the reader who has had some exposure to these ideas to think of the ways in which they could be useful in reliability analysis.

Much of the book relies on numerical techniques to fit complex models, and a natural question is how best to put these into practice. At one stage during the writing of the book, we were considering making a suite of programmes available on disk, but in the end we had to admit that the task of 'cleaning up' our programs to the extent that they could be used as a computer package would have held up the writing of the book even longer than the four years it has taken us to get this far; therefore we abandoned that project, and have contented ourselves with general advice, at appropriate points, about the algorithms to be used. Apart from Chapter 6, the main numerical technique is non-linear optimization, and there are numerous published subroutines for this, for example in the NAG or IMSL libraries or books such as Nash (1979) or Press et al. (1986). Our advice to the reader is to become familiar with one such subroutine, preferably one that does not require derivatives (NAG's E04CGF is an example) and to program up likelihood functions as required. An advantage of this, over providing a package to do everything in one go, is that it will force the reader to think about how numerical optimization techniques actually work, and thereby to pay attention to such things as appropriate scaling of the variables. The one exception to this is Chapter 6 on Bayes methods, since, except where conjugate priors are employed, this requires numerical integration. Some comments about exact and approximate methods for that are included in Chapter 6.

It is a pleasure to thank those who have helped us in the preparation of this book. Mike Bader of the University of Surrey and Leigh Phoenix of Cornell University have been valuable research collaborators, and the data in Tables 4.8.1 and 4.8.2 (based on experimental work by Mark Priest, at Surrey), 7.3.2 (from Bashear Gul-Mohammad) have come from the collaboration with Bader. Frank Gerstle gave us a copy of Gerstle and Kunz (1983) cited in sections 4.9, 4.10 and 5.4. Stephen M. Stigler of the University of Chicago drew our attention to the Buffon data of Section 4.10. Part of Chapter 4 is based on the paper by Smith (1991). Linda Wolstenholme, Karen Young and Leigh Phoenix provided us with very valuable comments on preliminary drafts of the manuscript. We also thank Sally Fenwick and Marion Harris for typing substantial portions of the manuscript, and Elizabeth Johnston of Chapman & Hall for her great expertise in bringing the project, despite many delays, to a successful conclusion. Finally, we gratefully acknowledge the Wolfson Foundation (R.L.S.) and the S.E.R.C. for financial support.

Martin J. Crowder
Alan C. Kimber
Richard L. Smith
Trevor J. Sweeting
Guildford.

1

Statistical concepts in reliability

1.1 INTRODUCTION

Reliability is a word with many different connotations. When applied to a human being, it usually refers to that person's ability to perform certain tasks according to a specified standard. By extension, the word is applied to a piece of equipment, or a component of a larger system, to mean the ability of that equipment or component to fulfil what is required of it. The original use of the term was purely qualitative. For instance, aerospace engineers recognized the desirability of having more than one engine on an aeroplane without any precise measurements of failure rate. As used today, however, reliability is almost always a *quantitative* concept, and this implies the need for methods of *measuring* reliability.

There are a number of reasons why reliability needs to be quantitative. Perhaps the most important is economic since to improve reliability costs money, and this can be justified only if the costs of unreliable equipment are measured. For a critical component whose successful operation is essential to a system, reliability may be measured as the probability that the component is operating successfully, and the expected cost of an unreliable component measured as the product of the probability of failure and the cost of failure. In a more routine application where components are allowed to fail but must then be repaired, the mean time between failures (MTBF) is a critical parameter. In either case, the need for a *probabilistic* definition of reliability is apparent.

Another reason for insisting on quantitative definitions of reliability is that different standards of reliability are required in different applications. In an unmanned satellite, for instance, a comparatively high failure rate may be considered acceptable, but in cases where human life is at stake, the probability of failure must be very close to zero. The applications where the very highest reliability are required are those in the nuclear industry, where it is common to insist on a bound of around 10^{-9} for the expected number of failures per reactor per year. Whether it is possible to guarantee such reliability in practice is another matter.

Whatever the specific application, however, reliability needs to be measured, and this implies the need for statistical methods. One well-known author on reliability has gone to considerable lengths to emphasize the distinction between physical quantities such as elastic modulus or electrical resistance, and reliability which by its nature is not a physical quantity. This is certainly an important distinction, but it does not mean that reliability cannot be measured, merely that the uncertainty of measurements is in most cases much higher than would be tolerated in a physical experiment. Consequently, it is important in estimating reliability to make a realistic appraisal of the uncertainty in any estimate quoted.

1.2 RELIABILITY DATA

This book is concerned primarily with methods for statistical analysis of data on reliability. The form of these data necessarily depends on the application being considered. The simplest case consists of a series of experimental units tested to a prescribed standard, and then classified as failures or survivors. The number of failures typically follows a Binomial or Hypergeometric distribution, from which it is possible to make inferences about the failure rate in the whole population.

More sophisticated applications usually involve a continuous measure of failure, such as failure load or failure time. This leads us to consider the *distribution* of the failure load or failure time, and hence to employ statistical techniques for estimating that distribution. There are a number of distinctions to be made here.

1. *Descriptive versus inferential statistics.*

    In some applications, it is sufficient to use simple measures such as the mean and variance, survivor function or hazard function, and to summarize these with a few descriptive statistics or graphs. In other cases, questions requiring more sophisticated methods arise, such as determining a confidence interval for the mean or a specified quantile of failure time, or testing a hypothesis about the failure-time distribution.

2. *Uncensored versus censored data.*

    It is customary to stop an experiment before all the units have failed, in which case only a lower bound is known for the failure load or failure time of the unfailed units. Such data are called right-censored. In other contexts, only an upper bound for the failure time may be known (left-censored) or it may be known only that failure occurred between two specified times (interval-censored data).

3. *Parametric versus nonparametric methods.*

    Many statistical methods take the form of fitting a parametric family such as the Normal, Lognormal or Weibull distribution. In such cases it is important to have an efficient method for estimating the parameters, but also to have ways of assessing the fit of a distribution. Other statistical procedures do not require any parametric form. For example, the Kaplan-

Meier estimator (section 2.11) is a method of estimating the distribution function of failure time, with data subject to right-censoring, without any assumption about a parametric family. At a more advanced level, proportional hazards analysis (Chapter 5) is an example of a *semiparametric* procedure, in which some variables (those defining the proportionality of two hazard functions) are assumed to follow a parametric family, but others (the baseline hazard function) are not.

4. *Single samples versus data with covariates.*

   Most texts on reliability concentrate on single samples – for example, suppose we have a sample of lifetimes of units tested under identical conditions, then one can compute the survival curve or hazard function for those units. In many contexts, however, there are additional explanatory variables or *covariates*. For example, there may be several different samples conducted with slightly different materials or under different stresses or different ambient conditions. This leads us into the study of reliability models which incorporate the covariates. Such techniques for the analysis of survival data are already very well established in the context of medical data, but for some reason are not nearly so well known among reliability practitioners. One of the aims of this book is to emphasize the possibilities of using such models for reliability data (especially in Chapters 4 and 5).

5. *Univariate and multivariate data.*

   Another kind of distinction concerns the actual variable being measured – is it a single lifetime and therefore a scalar, or are there a vector of observations, such as lifetimes of different components or failure stresses in different directions, relevant to the reliability of a single unit? In the latter case, we need multivariate models for failure data, a topic developed in Chapter 7.

6. *Classical versus Bayesian methods.*

   Historically, reliability and survival analysis have generally employed classical concepts of estimation, confidence intervals and hypothesis testing. During the last ten years or so, however, there has been increasing interest in Bayesian methods applied to reliability problems. A number of reasons might be identified for this. First, the rapid development of statistical computing, and in particular numerical integration routines for Bayesian statistics, have allowed Bayesian methods to be applied to complicated models of the sort employed in reliability theory. The paper of Naylor and Smith (1982), proposing new numerical integration routines which have since been extensively implemented, was particularly important in this development. Secondly, many reliability problems arise in situations where there is a high level of uncertainty, either because of the influence of uncertain factors which are hard to quantify, or because the reliability prediction problem involves significant extrapolation from any available data. Also, in many situations the data are simply inadequate for any predictions to be made with a high level of confidence. Thus it is recognized that practical judgements are very often, inevitably, strongly guided by subjective judgements. Bayesian statistics provides a means of quantifying subjective judgements and combining them in a rigorous way with information obtained from experimental data. Finally, there is vigorous debate over the philosophical basis of the Bayesian approach as it might be applied in reliability. The recent editorial by Evans (1989) and reply by Kaplan (1990) show that these issues are still very much alive. Bayesian methods form the subject of Chapter 6.

1.3 REPAIRABLE AND NONREPAIRABLE SYSTEMS

A rather more substantial distinction, which until recently was neglected in the reliability literature, is that between repairable and nonrepairable systems. However, in recent years publications such as the monograph of Ascher and Feingold (1984) and the Royal Statistical Society paper of Ansell and Phillips (1989) have emphasized that repairable systems need to be analyzed by different techniques from the more traditional survival data techniques used to analyze nonrepairable systems.

The following example illustrates a typical situation where it is necessary to be clear about the distinction. A new machine was placed into service at time 0, failed after 203 hours, and was replaced by another new machine. Subsequent failures and replacements occurred at cumulative times 286, 481, 873, 1177, 1438, 1852, 2091, 2295 and 2632 hours. The question is, what information does this give us about the failure rate and whether it is increasing or decreasing over the duration of the experiment?

Let us first consider a conventional analysis in which the inter-failure times are treated as a sample of independent, identically distributed failure times. The ordered failure times are $t = 83$, 195, 203, 204, 239, 261, 304, 337, 392 and 414 hours. A useful concept with data of this form is the **cumulative hazard function** (c.h.f.); precise definition of this is deferred to Chapter 2 but the intuitive notion is that an upwardly-curving c.h.f. indicates an increasing failure rate as the component ages, while a downwardly-curving c.h.f. indicates decreasing failure rate. A plot of the estimated $H(t)$ against t is given in Figure 1.1; the method of calculation is described in section 2.9. The plot is sharply increasing and convex over the range of interest. Since $H(t) = \int_0^t h(s)\, ds$ where h is the ordinary hazard function, which is often equated with the concept of a failure rate, we may conclude that the failure rate is increasing.

On the other hand, a simple plot of the proportion of failures up to (cumulative) time t, against t, yields Figure 1.2, which indicates a linear relationship. From this plot one would conclude that the failure rate is constant.

It should be clear here that we are dealing with different concepts of failure rate. The first concerns the failure rate of an individual item since the hazard rate is essentially trying to determine the probability that an item fails in a

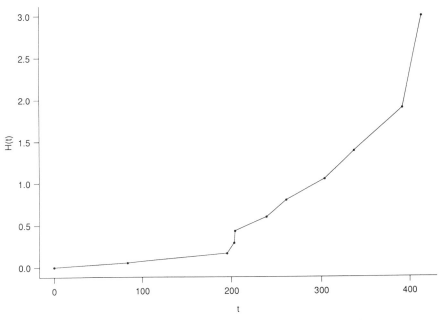

Figure 1.1 Cumulative hazard function of times between failure.

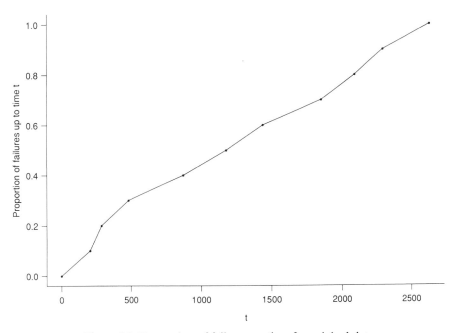

Figure 1.2 Proportion of failures *vs.* time for original data.

specified short interval given that it has survived up to the beginning of that interval. This measure is appropriate for a nonrepairable system in which it is the survival times of individual items which is of interest.

On the other hand, Figure 1.2 is essentially illustrating the behaviour of the system in real time and the fact that the plot is linear indicates that there is no change in the quality of the units over the time period being studied. This could be important if, say, there are changes in the manufacture of those units as time progresses, and is typically a question of interest in the analysis of repairable systems.

To underline the point further, suppose the units had actually been observed to fail in increasing time order, so that the cumulative times were 83, 278, 481, 685, 924, 1185, 1489, 1826, 2218 and 2632 hours. In this case Figure 1.1 remains unchanged, but Figure 1.2 becomes Figure 1.3. Now we can see a definite concavity, indicating that the failure rate is decreasing with time. In other words, the reliability of the system is improving. Presumably this is good news!

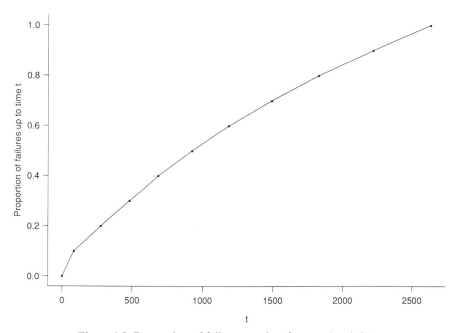

Figure 1.3 Proportion of failures *vs.* time for reordered data.

Thus we see a distinction between nonrepairable systems, in which it is the lifetimes of individual units that are of interest, and repairable systems, in which the point process of failure times is relevant. However, the distinction between repairable and nonrepairable systems is not as clear-cut as some authors have suggested. The hazard function of individual units is of interest even in a repairable system, and time trends, if present, are relevant whether a system is repairable or not. What is important is that there are different questions which need to be answered in different contexts, and any discussion of reliability concepts needs to be clear about these distinctions.

The reliability of repairable systems forms the subject of Chapter 8.

1.4 COMPONENT RELIABILITY AND SYSTEM RELIABILITY

For most of this book we talk of the reliability of *units* without trying to identify the place of those units in the overall system under study. In some contexts the unit will be the entire system, while in others it will be just one component in a much larger system.

The relation between component reliability and system reliability is of course very important, and many models have been widely studied. The simplest kinds of system are **series systems**, in which all the components must be working for the system to function properly, and **parallel systems** in which only one component is required to be working. These represent extremes of, in the one case, a system with no capacity to withstand failure and, in the other, one with a large amount of redundancy. These in turn suggest more complex systems, such as *k–out–of–n* systems, in which a system consisting of *n* components is able to function if at least *k* of the components are working correctly, and various systems which consist of combinations of series and parallel subsystems forming a kind of network structure. All these are examples of **coherent systems**, a subject for which an extensive literature has developed.

Coherent systems are not, however, the only kinds of systems for which it is possible to measure the relation between system and component reliability. One extension is the idea of **multi-state systems**, in which each component is classified into one of several states instead of being just failed or unfailed. Another kind of system is a **load-sharing system** in which the total load on the system is divided among the available components, with load redistribution if some of those components fail. Finally, a subject of very rapid recent development is that of software reliability, which has spawned some original probabilistic models of its own.

The bulk of this book being about statistics, we do not give very much attention to system reliability models. In particular, the algorithms which have been developed for reliability calculations in very large systems lie well beyond the scope of the book. We have, however, included in Chapter 9 a selective review of system reliability models. Particular emphasis is given there to load-sharing models as this is one subject which has not previously been covered in a book of this nature.

1.5 THE BINOMIAL AND HYPERGEOMETRIC DISTRIBUTIONS

We now turn our attention to some of the simpler statistical concepts used in reliability.

The Binomial and Hypergeometric distributions arise when units are classified into one of two groups, such as defective and nondefective. The *Binomial distribution* with parameters n and p is the probability distribution of the number X of defective (or, equivalently, nondefective) units in a sample of size n, when each one has an equal probability p of being defective (nondefective) and all units are independent. The formula is

$$\Pr\{X=k\}=\binom{n}{k}p^k(1-p)^{n-k}, \qquad 0\leq k\leq n. \tag{1.1}$$

The mean and variance of X are given by

$$\mathrm{E}\{X\}=np, \qquad \mathrm{Var}\{X\}=np(1-p). \tag{1.2}$$

In practice p is unknown, so we estimate p by $\hat{p}=X/n$ and the variance of X by $n\hat{p}(1-\hat{p})$. An approximate $100(1-\alpha)\%$ *confidence interval* for p is then

$$\left[\frac{X}{n}-z_{\alpha/2}\left(\frac{X(n-X)}{n^3}\right)^{1/2}, \frac{X}{n}+z_{\alpha/2}\left(\frac{X(n-X)}{n^3}\right)^{1/2}\right], \tag{1.3}$$

where $z_{\alpha/2}$ satisfies $1-\Phi(z_{\alpha/2})=\alpha/2$, Φ being the standard Normal distribution function. If the independence assumption is violated then the mean in (1.2) will still be correct but not the variance. In this case it will be necessary to estimate the variance by other means, for example by combining the results of several experiments. If there is doubt about the constancy of p, it may be necessary to look for trends in the data or to divide the data into subsamples and formally to test the equality of p. For more general problems, in which the probability of a defective unit depends on covariates, one must refer to texts on binary data, in particular Cox and Snell (1989).

If the sample is without replacement from a finite population, for example in sampling units from a large batch, then the *Hypergeometric distribution* replaces the binomial. This is given by

$$\Pr\{X=k\}=\binom{n}{k}\binom{N-n}{K-k}\bigg/\binom{N}{K}, \quad k=0,\ 1,\ 2,\ ... \tag{1.4}$$

where N is the population size and K the total number of defective units in the population. Writing $p = K/N$, equations (1.2) become

$$E\{X\} = np, \ \mathrm{Var}\{X\} = np(1-p)(N-n)/(N-1). \qquad (1.5)$$

Thus the principal effect of the finite population size is to reduce the variance by a factor $(N-n)/(N-1)$; this is called the **finite population correction**.

When n is large and $np > 5$, $n(1-p) > 5$, it is usual to approximate the Binomial distribution by a Normal distribution (Chapter 2) with mean $\mu = np$ and variance $\sigma^2 = np(1-p)$. The approximation (1.3) is based on this. For n large but $\lambda = np < 5$, a better approximation is via the Poisson distribution,

$$\Pr\{X = k\} \approx \lambda^k e^{-\lambda}/k!, \qquad k = 0, 1, 2, \ldots \qquad (1.6)$$

This has particular relevance as a model for the occurrence of very rare events. By symmetry, a Poisson approximation also holds for $n - X$ if $n(1-p) < 5$.

Example

1. A sample of size 30 is drawn from a very large population in which 5% of the items are defective. What is the probability that more than 10% of the items in the *sample* are defective?
2. Repeat for sample size 100.
3. Repeat the calculations of both (1) and (2) for the case in which sampling is without replacement for a total population of size 200.

(This example typifies the sort of calculations involved in sampling inspection or quality control schemes.)

Solution

1. We want the probability that there are more than 3 defective items in the sample. The probability of exactly 3 is $\{30!/(3!\ 27!)\}0.05^3 0.95^{27} = 0.12705$. Similarly, the probabilities of exactly 2, 1 or 0 defectives are 0.25864, 0.33890, 0.21464 so the probability that the number of defectives does not exceed 3 is 0.93923; the required answer is then $1 - 0.93923 = 0.06077$. *Alternatively*, we may use the Poisson distribution with mean 1.5; the desired probability is then $1 - 0.12551 - 0.25102 - 0.33469 - 0.22313 = 0.06565$. The two figures will be close enough for many practical purposes, but are not identical as can be seen.
2. The exact probability calculated from the Binomial distribution is 0.01147, and the Poisson approximation is 0.01370. In this case a Normal approximation with mean 5 and variance 4.75 does not give such good results: applying the *continuity correction*, the probability of more than 10 defectives is $1 - \Phi((10.5 - 5)/\sqrt{4.75})$ or $1 - \Phi(2.524) = 0.0058$, where Φ is the

standard Normal distribution function. The skewness of the Binomial distribution is causing problems here – compare the corresponding calculation that there are fewer than 0 defectives!

3. The hypergeometric probability of exactly 3 defectives is $\binom{30}{3}\binom{170}{7}/\binom{200}{10} = 0.12991$. Similarly, the probabilities for 2, 1 and 0 defectives are 0.28361, 0.35206 and 0.18894 leading to the answer 0.04548. For the sample size 100, since there are only 10 defectives in the entire population the probability of getting more than 10 in the sample is 0.

The preceding calculations were easily programmed in BASIC on a small microcomputer. For small samples, tables for the Binomial and Hypergeometric distributions exist. For example Lindley and Scott (1984) tabulate the Binomial distribution for samples up to size 20 and the Hypergeometric for samples up to size 17.

1.6 THE POISSON PROCESS

An alternative derivation of the Poisson distribution (1.6) is via a special model for the times of random events, the **Poisson process**.

Suppose we are observing a series of random events. For definiteness, we suppose that the events are failures of units, so that the observations are of failure times, perhaps in a repairable system. Natural assumptions, which may or may not be satisfied in any particular example, are:

1. failures occurring in disjoint time intervals are (statistically) independent;
2. the failure rate (mean number of failures per unit time) is constant, and so does not depend on the particular time interval examined.

If both assumptions are satisfied, then the process is called a Poisson process with (failure) rate λ say. Two important properties of a Poisson process are:

1. the number of failures X in a time interval of length t has a Poisson distribution with mean λt, so that

$$\Pr\{X = k\} = (\lambda t)^k e^{-\lambda t}/k!, \; k \geq 0;$$

2. the times between successive failures are independent random variables, each having an exponential density with parameter λ, so that

$$\Pr\{\text{Failure time} > x\} = e^{-\lambda x}, \qquad 0 < x < \infty.$$

The mean time between failures (MTBF) is λ^{-1}.

The first property is closely related to equation (1.6). Indeed, one kind of system for which we would expect the Poisson process to be a good model is one in which there are very many components liable to failure, but the probability of failure of any one of them is small. This corresponds to the 'rare events' interpretation of equation (1.6). The second property suggests the exponential distribution as a model for *survival times*.

In applications, assumption (2) may be critical. Many systems are in practice either improving or deteriorating with time. In this case, we shall need more general models. The simplest of these is a **nonhomogeneous** Poisson process in which the failure rate is non-constant. This model is particularly important in the analysis of repairable systems (Chapter 8).

1.7 THE RELIABILITY LITERATURE

Numerous books are available presenting reliability at various levels of mathematical sophistication. Books such as Greene and Bourne (1972) and O'Connor (1985) are aimed primarily at engineers, presenting statistical techniques at an introductory level and emphasizing their practical application. At the other end of the spectrum lie mathematical texts such as Barlow and Proschan (1981), who give detailed coverage of the probabilistic models relating component to system reliability. In between, and closer to the spirit of the present book, lie a number of texts on survival data analysis, some oriented more towards engineering applications and others towards medical data. In the former category lie Mann, Schafer and Singpurwalla (1974) and Nelson (1982), while Lawless (1982) and Cox and Oakes (1984) are more in the latter category. All these books, however, give good coverage of the concepts and techniques used in analyzing reliability data. Kalbfleisch and Prentice (1980) is an advanced text with particular emphasis on proportional hazards methods, while Martz and Waller (1982) is the only existing work to emphasize the Bayesian approach to reliability. Repairable systems reliability is covered by the book of Ascher and Feingold (1984), while Thompson (1988) covers point process models, a topic particularly relevant in analyzing repairable systems. Other books relevant to point processes are Cox and Lewis (1966), which concentrates on statistical aspects, and Cox and Isham (1980), which covers probabilistic theory.

In addition to books on reliability, there are several journals and conference proceedings. *IEEE Transactions on Reliability* is, as its name implies, a journal devoted to reliability, as is *Reliability Engineering and System Safety*. Also, several of the major statistics and OR journals publish reliability papers; these include *Technometrics, Applied Statistics* and *Operations Research*. Conference proceedings include the *Reliability* series sponsored by the United Kingdom Atomic Energy Authority, and there have also been numerous specific volumes or special issues of journals, for example the Weibull memorial volume (Eggwertz and Lind, 1985) and the special issue of *Operations Research* (1984, **32** (3)). Finally, we would draw the reader's attention to the discussion papers of Bergman (1985) and Ansell and Phillips (1989) for a broad-ranging discussion of statistical issues relating to reliability.

2

Probability distributions in reliability

2.1 INTRODUCTION

In many areas of statistical application, the natural starting point to model a random variable of interest is the Normal distribution. This may result from purely pragmatic considerations or from a theoretical argument based on the Central Limit Theorem, which says that if the random variable is the sum of a large number of small effects, then its distribution is approximately Normal. In the context of reliability, the case for Normality is much less compelling. For one thing, lifetimes and strengths are inherently positive quantities. In addition, from a modelling point of view, it is perhaps natural to begin with the Poisson process ideas discussed in section 1.6, which lead to the exponential distribution. Whilst this distribution is of limited applicability in practice, generalizations of the exponential distribution, such as the gamma and Weibull distributions have proved to be valuable models in reliability. These and other probability distributions commonly encountered in reliability are discussed in this chapter in sections 2.3 to 2.7.

Other distinctive aspects of the statistical analysis of reliability data are the central role played by the survivor and hazard functions, and the natural occurrence of censored observations. These issues are discussed here in sections 2.2 and 2.8.

Finally in this chapter we put the probabilistic results in the context of analyzing data. Whilst general statistical methods for fitting probability distributions to data and for assessing their goodness of fit are covered in detail in Chapter 3, we introduce some simple statistical methods in sections 2.9 to 2.11.

2.2 PRELIMINARIES ON LIFE DISTRIBUTIONS

To fix ideas, let T be the random time to failure of a unit under study. Here we use time in its most general sense. It may be real time or operational time or indeed any non-negative variable, such as breaking strain or number of revolutions to failure. Let

$$F(t) = \Pr(T < t)$$

be the **distribution function** of T and let

$$S(t) = \Pr(T \geq t) = 1 - F(t)$$

be the **survivor function** or **reliability function** of T. Note that some authors define $F(t)$ and $S(t)$ by $\Pr(T \leq t)$ and $\Pr(T > t)$ respectively. In practice this makes no difference to the results that follow when T is a continuous random variable, which is the case we shall mainly consider from now on. We shall assume that T has a **density function**

$$f(t) = \frac{dF(t)}{dt} = -\frac{dS(t)}{dt},$$

so that the probability that a unit fails in the short time interval $[t, t+\delta t)$ is

$$\Pr(t \leq T < t+\delta t) \cong f(t)\delta t.$$

Consider now the same event, that is $t \leq T < t+\delta t$, but this time conditional on the fact that the unit has not failed by time t. That is

$$\Pr(t \leq T < t+\delta t \,|\, T \geq t) \cong \frac{f(t)\delta t}{S(t)}.$$

This may be thought of as the probability of imminent failure at time t. The function $h(t)$ given by

$$h(t) = f(t)/S(t)$$

is the **hazard** or **failure rate function**, and is a natural indicator of the 'proneness to failure' of a unit after time t has elapsed. The **cumulative hazard function** is

$$H(t) = \int_0^t h(u)\, du,$$

and it is readily seen that

$$S(t) = \exp\{-H(t)\}. \tag{2.1}$$

Note that f, F, S, h and H give mathematically equivalent descriptions of T in the sense that, given any one of these functions the other four functions may be deduced.

Some typical cases are briefly discussed below.

1. If $h(t)=\lambda$, a constant, then $H(t)=\lambda t$ and $S(t)=\exp(-\lambda t)$, the survivor function of the exponential distribution with rate parameter λ. The corresponding density function is

$$f(t)=\lambda e^{-\lambda t}.$$

Thus, an exponential life distribution corresponds to no ageing, and is a natural starting point in reliability modelling.

2. If $h(t)$ is an increasing function of t, then T is said to have an increasing failure rate (IFR). This is appropriate when the unit is subject to ageing, perhaps through wear, fatigue or cumulative damage.

3. If $h(t)$ is a decreasing function of t, then T is said to have a decreasing failure rate (DFR). This may occur, for example, when a manufacturing process produces an appreciable proportion of low quality units which are likely to fail early. After a while, higher quality components remain, which have lower failure rates. This is a common situation with some electronic devices. In such cases, initial proof-testing of units is often carried out in order to eliminate the substandard units (so-called 'burn-in').

4. Another commonly mentioned case is the bath-tub hazard, which has decreasing hazard initially but eventually has increasing hazard. A scenario in which one might observe this type of hazard is as follows. Substandard units tend to fail early, leaving higher quality components. These will tend to have a low and flat hazard for some design life, after which time fatigue becomes increasingly a factor, causing a steady increase in hazard.

It is customary in textbooks on reliability and survival analysis (including this one) to mention the bath-tub hazard early on, explain how it might arise, and then largely to ignore it for the rest of the text. This is justified on two grounds. First, the relative intractability of fitting probability models with bath-tub hazard is a factor. Second, it is argued that in most situations of practical interest, the very weak units are screened out (perhaps as a result of quality control) early and before the observation period begins. Thus the decreasing part of the hazard is absent. Some interesting results in 'hazard sculpting' may be found in Gaver and Acar (1979).

2.3 THE EXPONENTIAL DISTRIBUTION

As mentioned above and in section 1.6, the exponential distribution is a natural starting point as a reliability distribution. To reiterate, the exponential distribution has survivor, hazard and density functions given by

$$S(t) = \exp(-\lambda t)$$
$$h(t) = \lambda \tag{2.2}$$
$$f(t) = \lambda \exp(-\lambda t),$$

where λ is a positive parameter, often called the rate parameter, and where $t > 0$. Note also that the exponential distribution has mean $1/\lambda$, variance $1/\lambda^2$ and is positively skewed. The shape of the density is the same for all λ, and $1/\lambda$ acts as a scale parameter. So, for example, if the lifetime in minutes, T, of a certain component is exponentially distributed with rate parameter λ, its lifetime in hours is $T^* = T/60$ and T^* is exponentially distributed with rate parameter 60λ. A commonly encountered alternative formulation is to parametrize the distribution via $\alpha = 1/\lambda$ in place of λ. Figure 2.1 shows two exponential densities.

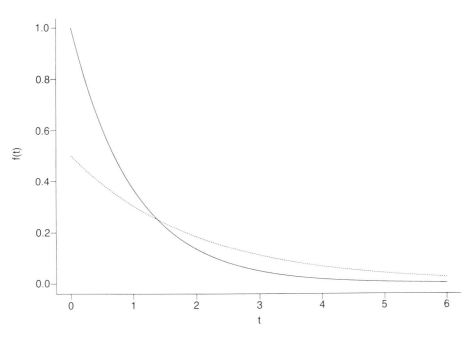

Figure 2.1 Exponential density functions with mean 1 (solid) and mean 2 (dotted).

The corresponding hazard functions are given in Figure 2.2. We shall see below that the exponential distribution is a special case of both the Weibull and gamma families of distributions.

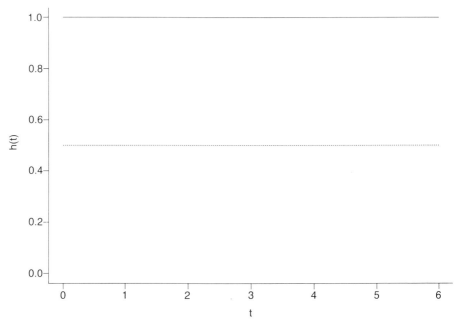

Figure 2.2 Exponential hazard functions with mean 1 (solid) and mean 2 (dotted).

2.4 THE WEIBULL AND GUMBEL DISTRIBUTIONS

A Weibull random variable (after W. Weibull (1939, 1951)) is one with survivor function

$$S(t) = \exp\{-(t/\alpha)^\eta\}, \tag{2.3}$$

for $t > 0$ and where α and η are positive parameters, α being a scale parameter and η being a shape parameter. Note that when $\eta = 1$, we obtain an exponential distribution with $\lambda = 1/\alpha$.

The Weibull hazard function is

$$h(t) = \eta \alpha^{-\eta} t^{\eta-1}.$$

This is DFR for $\eta < 1$, constant for $\eta = 1$ (exponential) and IFR for $\eta > 1$. In particular, for $1 < \eta < 2$, the hazard function increases slower than linearly; for $\eta = 2$ the hazard function is linear; and for $\eta > 2$ the hazard increases faster than linearly. A selection of Weibull hazard functions is shown in Figure 2.3.

The Weibull density is

$$f(t) = \eta \alpha^{-\eta} t^{\eta-1} \exp\{-(t/\alpha)^\eta\}, \tag{2.4}$$

for $t>0$. The mean and variance are given by $\alpha\Gamma(\eta^{-1}+1)$ and $\alpha^2\{\Gamma(2\eta^{-1}+1)-[\Gamma(\eta^{-1}+1)]^2\}$, where Γ is the gamma function

$$\Gamma(x)=\int_0^\infty u^{x-1}e^{-u}\,du, \tag{2.5}$$

see, for example, Abramowitz and Stegun (1972, Chapter 6). A Fortran program for computing equation (2.5) is given in Griffiths and Hill (1985, pp. 243–6), which is based on an earlier program of Pike and Hill (1966). When η is large (greater than 5, say), the mean and variance are approximately α and $1.64\alpha^2/\eta^2$ respectively. The shape of the density depends on η. Some Weibull densities are shown in Figure 2.4.

The Weibull distribution is probably the most widely used distribution in reliability analysis. It has been found to provide a reasonable model for lifetimes of many types of unit, such as vacuum tubes, ball bearings and composite materials. A possible explanation for its appropriateness rests on its being an extreme value distribution; see Galambos (1978). Moreover, the closed form of the Weibull survivor function and the wide variety of shapes exhibited by Weibull density functions make it a particularly convenient generalization of the exponential distribution.

The Gumbel (or extreme-value, or Gompertz) distribution has survivor function

$$S(x)=\exp\{-\exp[(x-\mu)/\sigma]\} \tag{2.6}$$

for $-\infty<x<\infty$, where μ is a location parameter and $\sigma>0$ is a scale parameter. This distribution also arises as one of the possible limiting distributions of minima, see Galambos (1978), and has exponentially increasing failure rate. It is sometimes used as a lifetime distribution even though it allows negative values with positive probability. More commonly, however, the Gumbel distribution arises as the distribution of $\log T$. This is equivalent to assuming that T has a Weibull distribution. The relationship between the Gumbel and Weibull parameters is $\mu=\log\alpha$ and $\sigma=1/\eta$.

The Gumbel density function is

$$f(x)=\sigma^{-1}\exp\{(x-\mu)/\sigma\}S(x) \tag{2.7}$$

for $-\infty<x<\infty$, and has the same shape for all parameters. Note that the mean and variance of a Gumbel random variable are $\mu-\gamma\sigma$ and $(\pi^2/6)\sigma^2$ respectively, where $\gamma=0.5772\ldots$ is Euler's constant, and the distribution is negatively skewed. The density and hazard functions for a Gumbel distribution with $\mu=0$ and $\sigma=1$ are shown in Figures 2.5 and 2.6 respectively.

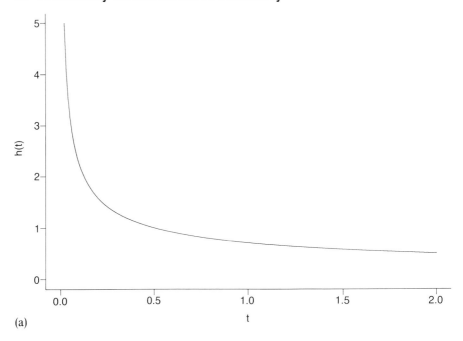

(a)

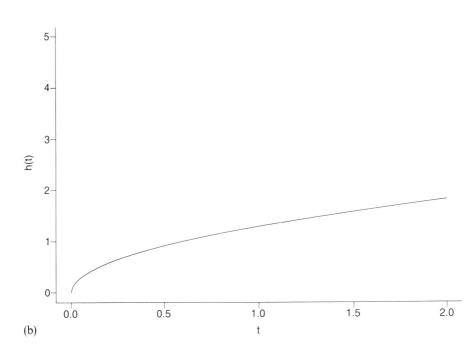

(b)

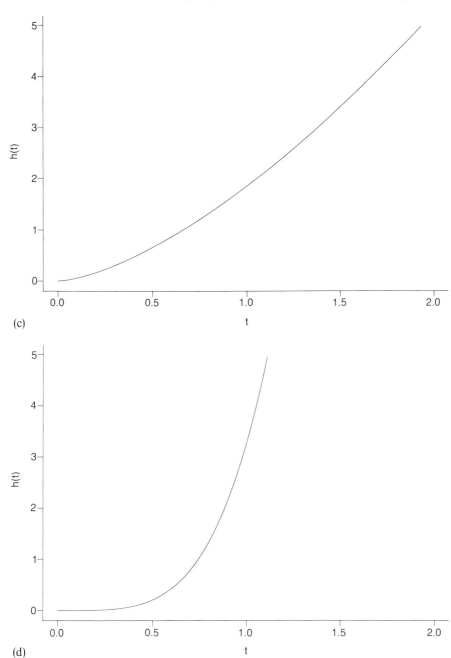

(c)

(d)

Figure 2.3 Hazard functions for four Weibull distributions with mean 1 and (a) $\eta=0.5$, (b) $\eta=1.5$, (c) $\eta=2.5$ and (d) $\eta=5$.

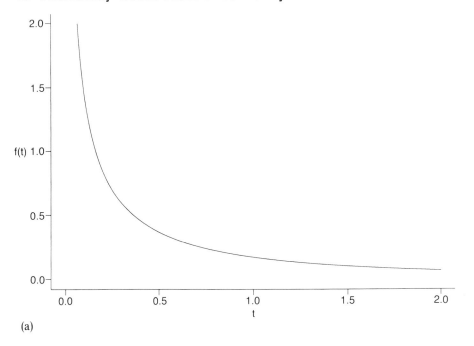

(a)

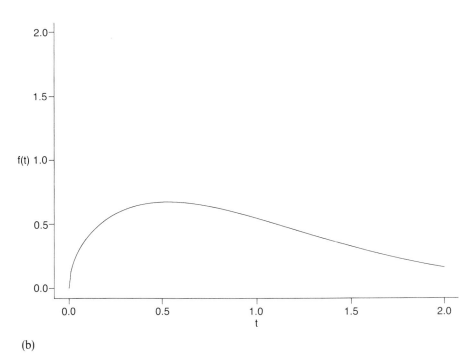

(b)

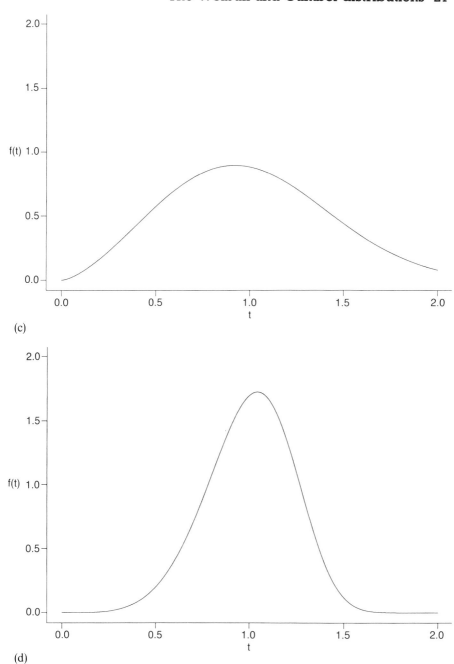

(c)

(d)

Figure 2.4 Density functions for four Weibull distributions with mean 1 and (a) $\eta = 0.5$, (b) $\eta = 1.5$, (c) $\eta = 2.5$ and (d) $\eta = 5$.

2.5 THE NORMAL AND LOGNORMAL DISTRIBUTIONS

The most commonly used distribution in statistics is the Normal distribution with density function

$$f(x)=(2\pi\sigma^2)^{-1/2}\exp\{-(x-\mu)^2/(2\sigma^2)\}$$

for $-\infty<x<\infty$, with mean μ and variance σ^2. When $\mu=0$ and $\sigma=1$ we have

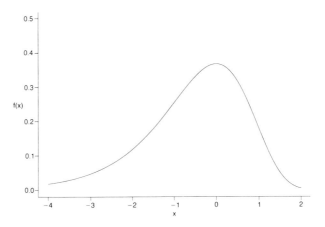

Figure 2.5 Density function of the Gumbel distribution with $\mu=0$ and $\sigma=1$.

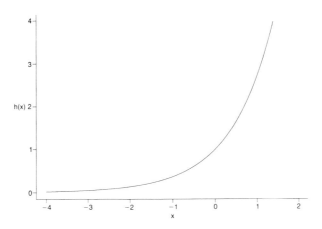

Figure 2.6 Hazard function of the Gumbel distribution with $\mu=0$ and $\sigma=1$.

the standard Normal distribution, the density and hazard functions for which are shown in Figures 2.7 and 2.8 respectively.

The Normal distribution is sometimes used as a lifetime distribution, even though it allows negative values with positive probability. More frequently,

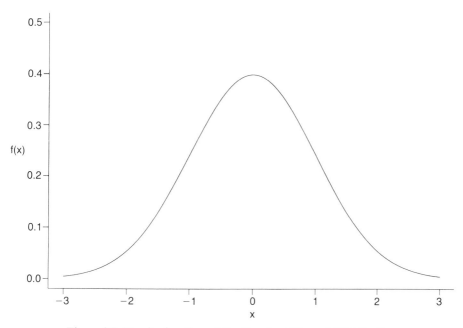

Figure 2.7 Density function of the Standard Normal Distribution.

however, it is used as a model for log T, the log-lifetime. This is equivalent to assuming a lognormal distribution for the lifetimes. The lognormal density is given by

$$f(t) = (2\pi\sigma^2 t^2)^{-1/2} \exp\{-(\log t - \mu)^2/(2\sigma^2)\}$$

for $t > 0$, where μ and σ are as in the normal distribution. The mean and variance of the lognormal distribution are $\exp(\mu + \tfrac{1}{2}\sigma^2)$ and $\exp(2\mu + \sigma^2)$ $\{\exp(\sigma^2) - 1\}$ respectively. Some examples of lognormal densities are shown in Figure 2.9. Note that for small σ, the lognormal density looks very like a Normal density. Some theoretical justification for using a Normal or lognormal distribution comes from the Central Limit Theorem when T or log T can be thought of as the sum of a large number of small effects.

The survivor and hazard functions of Normal and lognormal distributions can only be given in terms of integrals. The hazard functions of some

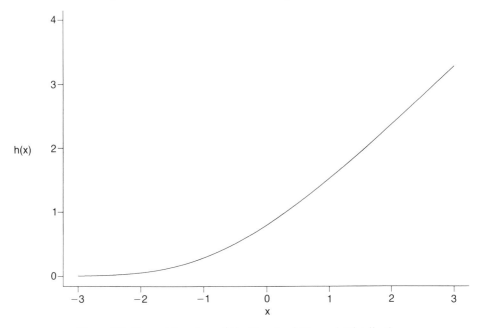

Figure 2.8 Hazard function of the Standard Normal Distribution.

lognormal distributions are given in Figure 2.10. Note that they are initially increasing, but eventually decreasing, approaching zero as $t \to \infty$. This behaviour is somewhat counter to what one might expect of lifetimes in practice, though it does not rule out the lognormal as a pragmatic fit for lifetimes when predictions for long lifetimes are not of interest.

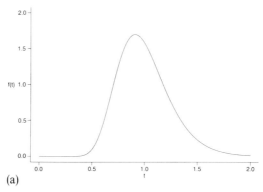

(a)

Figure 2.9 Density functions for four lognormal distributions with mean 1 and (a) $\sigma = 0.25$, (b) $\sigma = 0.5$, (c) $\sigma = 1$ and (d) $\sigma = 1.25$.

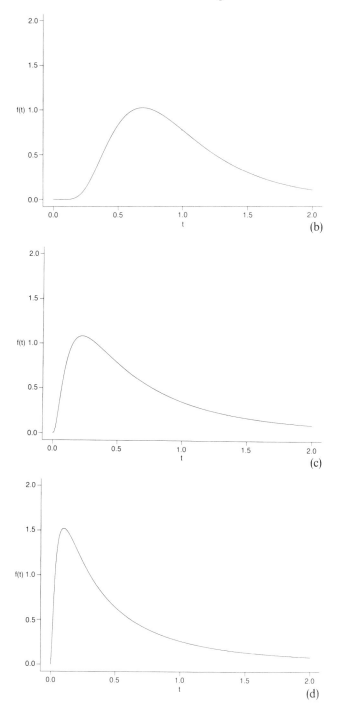

(b)

(c)

(d)

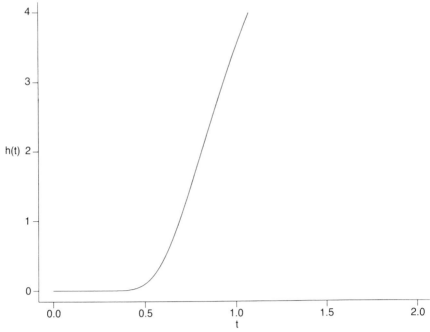

(a)

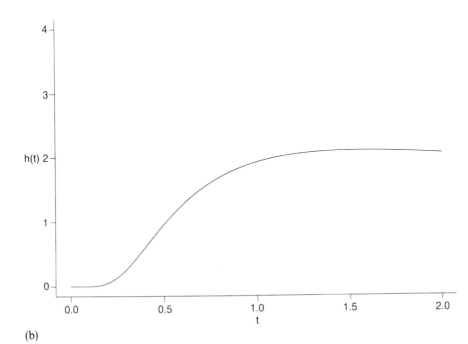

(b)

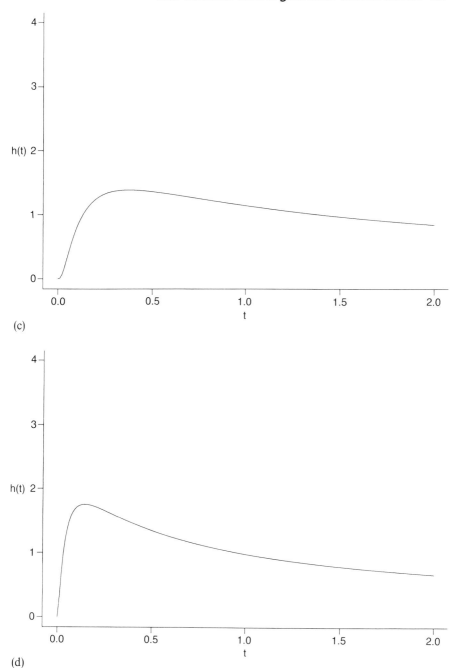

(c)

(d)

Figure 2.10 Hazard functions for four lognormal distributions with mean 1 and (a) $\sigma = 0.25$, (b) $\sigma = 0.5$, (c) $\sigma = 1$ and (d) $\sigma = 1.25$.

2.6 THE GAMMA DISTRIBUTION

The gamma distribution with parameters $\lambda > 0$ and $\rho > 0$ has density function

$$f(t) = \frac{\lambda^\rho t^{\rho - 1} \exp(-\lambda t)}{\Gamma(\rho)} \tag{2.8}$$

for $t > 0$ and where $\Gamma(.)$ is the gamma function given in (2.5). The mean and variance are ρ/λ and ρ/λ^2, $1/\lambda$ is a scale parameter and ρ is a shape parameter. Note that $\rho = 1$ gives an exponential distribution with rate parameter λ. Gamma distributions are positively skewed, though the skewness tends to zero for large ρ, in which case the density resembles a Normal density. Four gamma densities are shown in Figure 2.11. The gamma survivor function in general cannot be written in closed form, although for integer ρ, we have

$$S(t) = e^{-\lambda t} \left\{ 1 + (\lambda t) + \frac{(\lambda t)^2}{2!} + \cdots + \frac{(\lambda t)^{\rho - 1}}{(\rho - 1)!} \right\}. \tag{2.9}$$

The hazard function is either DFR for $\rho < 1$, flat in the exponential case with $\rho = 1$, or IFR for $\rho > 1$.

Some gamma hazard functions are given in Figure 2.12. A possible justification for using the gamma distribution is, at least when ρ is an integer, if the lifetime of a unit may be regarded as the time to ρth failure in a Poisson process. This

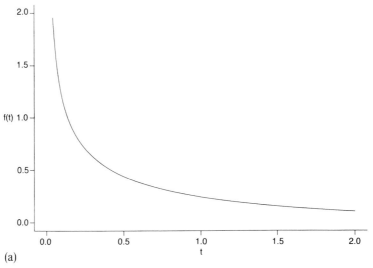

(a)

Figure 2.11 Density functions for four Gamma distributions with mean 1 and (a) $\rho = 0.5$, (b) $\rho = 1.5$, (c) $\rho = 2.5$ and (d) $\rho = 5$.

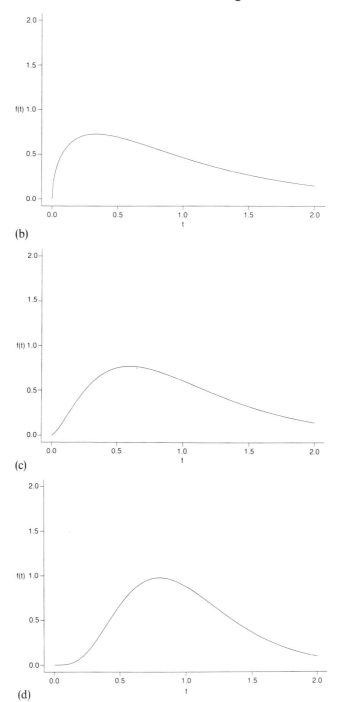

(b)

(c)

(d)

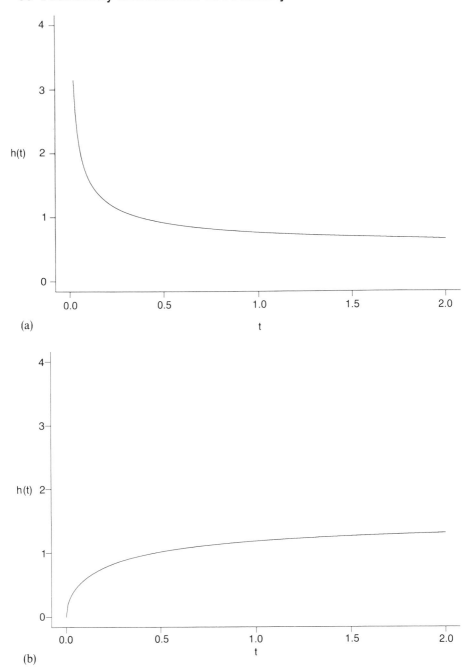

Figure 2.12 Hazard functions for four Gamma distributions with mean 1 and (a) $\rho = 0.5$, (b) $\rho = 1.5$, (c) $\rho = 2.5$ and (d) $\rho = 5$.

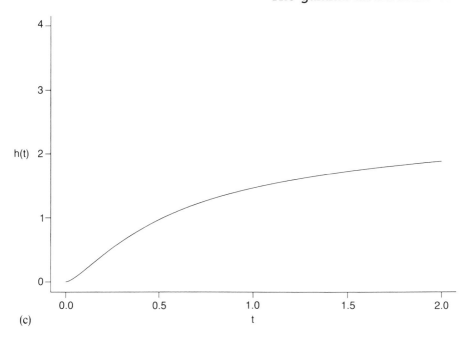

(c)

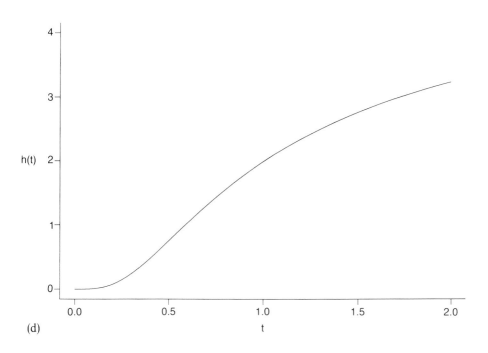

(d)

might be the case if a unit consists of ρ independent subunits, $\rho - 1$ of which are redundant, each of which has exponential lifetime.

Since the exponential distribution is a special case of the gamma, the gamma distribution may be regarded as a generalization of the exponential and an alternative to the Weibull distribution.

In the Weibull and lognormal models for T, it is sometimes more convenient to work with $\log T$ since the resulting Gumbel and Normal distributions depend only on location and scale parameters. However, if T is gamma, then $\log T$ has what is sometimes called the log-gamma distribution, which still involves a shape parameter. Thus, it is as convenient to work with T and the gamma distribution as it is to transform to a log-scale.

2.7 SOME OTHER LIFETIME DISTRIBUTIONS

In the previous sections we have discussed the most commonly used one- and two-parameter models for lifetime distributions. In this section we examine briefly some further possibilities, all of which introduce additional parameters in order to allow greater flexibility or to describe the mechanism of failure more fully. The cost of this increase in flexibility or descriptive power is, of course, greater complexity.

Guarantee parameter
One approach is to introduce a guarantee, or threshold, or shift parameter into the framework. This introduces a location parameter θ and implies that all lifetimes must exceed θ. A simple example is the shifted exponential distribution with density

$$f(t) = \lambda \exp\{-\lambda(t-\theta)\}$$

for $t > \theta$. Another example is the shifted Weibull or three-parameter Weibull distribution with density

$$f(t) = \eta \alpha^{-\eta}(t-\theta)^{\eta-1} \exp\{-[(t-\theta)/\alpha]^\eta\}, \qquad (2.10)$$

for $t > \theta$.

In general we replace t by $t - \theta$ in the right-hand side of the density expression. Taking $\theta = 0$ gives the unshifted version of the density. In the reliability context one would usually expect θ to be non-negative. If θ is known then there are no new problems since all that is being done is to work with $T - \theta$ in place of T itself. However, if θ is unknown we shall see in Chapter 3 that complications may occur.

Mixtures
A second generalization is via mixtures. Suppose that the population consists of two types of unit, one with lifetime density f_1 and the other with lifetime density f_2. For example, one type of unit may be inherently weaker than the

other type. Thus, the mean lifetime corresponding to density f_1 may be much lower than that corresponding to density f_2. If p is the proportion of type one units in the population, then the density of a randomly selected lifetime is

$$f(t) = pf_1(t) + (1-p)f_2(t),$$

for $t > 0$. For example, if f_i is an exponential density with rate parameter $\lambda_i (i = 1, 2)$, then the two-component mixture density is

$$f(t) = p\lambda_1 \exp(-\lambda_1 t) + (1-p)\lambda_2 \exp(-\lambda_2 t)$$

for $t > 0$ and $0 < p < 1$, giving a three-parameter model. Of course, one can extend this argument to any finite number of subdivisions of the population. Thus, for example, a k-component exponential mixture will involve $2k-1$ parameters and a k-component Weibull mixture will involve $3k-1$ parameters. Mixtures of this type can give rise to density functions with two or more modes (that is, two or more local maxima) and non-monotonic hazard functions. Useful references are Titterington, Smith and Makov (1985) and Everitt and Hand (1981).

An alternative approach is to suppose that each unit has a lifetime from a particular probability model, but that some or all of the parameters of the model vary over the population of units according to some continuous probability distribution. An interesting and tractable three-parameter model arises in this way by considering a Weibull–gamma mixture as follows.

Suppose, given α, that T is Weibull distributed with parameters η and α. Thus, conditionally on α, T has survivor function (2.3). Then if $\alpha^{-\eta}$ has a gamma distribution with parameters ρ and λ, and density (2.8), unconditionally T has survivor function

$$S(t) = \frac{\lambda}{(\lambda + t^\eta)^\rho}.$$

In this case T is said to have a Burr distribution; see Burr (1942) and Crowder (1985). When $\rho = 1$ this is sometimes called the log-logistic distribution.

The hazard function is

$$h(t) = \frac{\rho \eta t^{\eta - 1}}{(\lambda + t^\eta)}.$$

Note that this hazard function is DFR when $\eta \le 1$ and non-monotonic when $\eta > 1$.

Models derived from multiple modes
Suppose now that there are two failure modes, A and B and that the lifetime of a unit if **only** failure mode A (or B) is acting is T_A (or T_B). Then the actual

lifetime of a unit is

$$T = \min(T_A, T_B).$$

If the mode of failure is known, then we are in a competing risks framework; see Cox and Oakes (1984, Chapter 9) and Chapter 7. However, if the failure mode is unknown, only T is observable. In this case and if T_A and T_B are independent with survivor functions S_A and S_B respectively, then the survivor function of a randomly selected lifetime is

$$S(t) = S_A(t)S_B(t). \tag{2.11}$$

For example, if T_A is exponential with rate parameter λ_1 and T_B independently is gamma with $\rho = 2$ and $\lambda = \lambda_2$, then using equations (2.2), (2.9) and (2.11) gives

$$S(t) = \exp(-\lambda_1 t)(1 + \lambda_2 t) \exp(-\lambda_2 t)$$
$$= (1 + \lambda_2 t) \exp\{-(\lambda_1 + \lambda_2)t\}.$$

This approach may be generalized to cope with any number of failure modes. Similar ideas have been used in hydrology, but involving maxima rather than minima; see, for example, Rossi *et al.* (1984).

Generalized gamma distribution
We have seen earlier that the exponential distribution is a special case of both the gamma and Weibull families of distributions. In order to create a richer family of distributions that includes the gamma and Weibull distributions, the generalized gamma distribution has been proposed; Stacy (1962). The density is given by

$$f(t) = \frac{\eta \lambda (\lambda t)^{\rho \eta - 1} \exp\{-(\lambda t)^\eta\}}{\Gamma(\rho)}$$

for $t > 0$, where η and ρ are shape parameters and $1/\lambda$ is a scale parameter. Some important special cases are as follows:

Exponential	$\eta = \rho = 1$
Weibull	$\rho = 1$
Gamma	$\eta = 1$

Generalized extreme value distribution
Another generalization of the Weibull model is the generalized extreme value distribution. This may be used as a model for T or for log T. In the first case

the survivor function for T is

$$S(t) = \exp\left\{-\left[1 - \gamma\left(\frac{t-\mu}{\sigma}\right)\right]^{-1/\gamma}\right\},$$

where μ, $\sigma > 0$ and $\gamma < 0$ are parameters and $t > (\sigma/\gamma) + \mu$. This is a sometimes convenient reparametrization of the three-parameter Weibull distribution with density (2.10); see Smith and Naylor (1987). The limiting case as $\gamma \to 0$ is the Gumbel distribution with survivor function (2.6).

When used as a model for $\log T$ the generalized extreme value distribution leads to the following survivor function for T.

$$S(t) = \exp\left\{-\left[1 - \gamma\left(\frac{\log t - \mu}{\sigma}\right)\right]^{-1/\gamma}\right\} \tag{2.12}$$

where $t > 0$, and where μ, $\sigma > 0$ and $\gamma < 0$ are parameters. The limiting case as $\gamma \to 0$ is equivalent to the Weibull distribution and (2.12) reduces to (2.3) with $\mu = \log \alpha$ and $\sigma = 1/\eta$.

Hazard-based models
Here we begin by giving the hazard function an appropriate structure, from which we may deduce other properties such as the survivor function. For example, a bath-tub hazard could be modelled as a quadratic function

$$h(t) = a + 2bt + 3ct^2$$

for appropriate values of a, b and c (to give a bath-tub shape and ensure that $h(t) \geq 0$ for all t). Thus, the cumulative hazard is

$$H(t) = at + bt^2 + ct^3,$$

giving survivor function

$$S(t) = \exp(-at - bt^2 - ct^3)$$

and density function

$$f(t) = (a + 2bt + 3ct^2) \exp(-at - bt^2 - ct^3).$$

See Gaver and Acar (1979) for a discussion of this approach.

Many other statistical models have been applied to reliability data, though the above models are the most commonly encountered in practice. Other models that are sometimes applied to reliability data are the inverse Gaussian

(Chhikara and Folks, 1977 and Whitmore, 1983) and the Normal distribution truncated on the left at zero (Davis, 1952).

2.8 CENSORED DATA

Life data frequently have the complicating feature of containing 'incomplete' observations. This commonly occurs when the exact lifetime of a unit is not observed, but it is known to exceed a certain time, c say. Such an observation is referred to as **right-censored**. A right-censored lifetime for a unit might arise, for example, if the unit is still 'alive' or in operation at the end of the time period set aside for observation.

A **left-censored** observation is one in which the unit is known only to have failed prior to some time c. This situation might arise if a unit were put on test, but only checked for failure every hour. If, at the first check after one hour, the unit is found to have failed, then one knows only that its lifetime was less than one hour. If, in this scenario, the unit was found to have failed between the second and third checks (that is, the unit was working at the second check, but had failed by the third check) then one would know that the unit had lifetime of between two and three hours. This is an example of **interval censoring**.

We shall see later that censored observations may be dealt with, at least in principle, in a straightforward way provided that the mechanism of censoring is independent of the lifetimes of the units. Thus, one possible censoring mechanism, known as Type I censoring, is to put n units on test and observe them for a pre-arranged fixed time period c. Thus, the ith lifetime T_i ($i = 1, 2, \ldots, n$) is observed if $T_i \leq c$, otherwise it is known only that $T_i > c$. This type of censoring can be handled easily. However, if the censoring mechanism were as follows, then the statistical methods featured later would be inappropriate, and a more complicated methodology would be needed. Suppose that n units were put on test and that each unit was observed until the observer was sure that the unit was beginning to fail. If the observer's view is based on some knowledge and is not pure guesswork, then the censoring mechanism itself contains information about the lifetimes of interest. For example, under this mechanism if T_i is right-censored at c_i, we know more than the fact that $T_i > c_i$: we know that $T_i = c_i + \varepsilon_i$, where ε_i is some positive random variable with 'small' mean and variance (the smallness of which depends on the expertise of the observer). Thus, under this scheme a unit censored at c_i would not be representative of all those units that had survived c_i time units or more.

A common form of right-censoring in reliability studies is Type II censoring. Here observation ceases after a pre-specified number of failures. Note that the right-censoring time (or times if not all units are put on test at the same start time) is (are) not known in advance. Other more complicated forms of right-censoring may arise, but these may be handled routinely providing that any unit whose lifetime is right-censored at c_i is representative of all similar units whose lifetimes exceed c_i. Similar remarks apply to left- and interval-censoring.

2.9 SIMPLE DATA ANALYTIC METHODS: NO CENSORING

In this, and the remaining sections of the chapter we introduce some simple methods of analyzing (at least informally) component reliability data. Whilst more formal methods will be considered in Chapter 3, here the intention is to derive and display the data-based analogues of some of the quantities discussed in section 2.2 in relation to some of the models featured in sections 2.3 to 2.7.

To fix ideas, suppose that n units have been put on test and their lifetimes $t_1, t_2, \ldots, t_n$ have been observed. For the moment we shall assume that there has been no censoring.

In such a simple situation we can use standard elementary methods to describe the data and perhaps check on the plausibility of some particular models. For example, we may construct the histogram, which may be regarded as a crude sample analogue of the density function, calculate sample moments and other summary measures, such as the median and the quartiles. For certain simple models we may be able to exploit relationships between moments to check on the plausibility of the models and perhaps obtain rough parameter estimates.

Example 2.1
The following data from Lieblein and Zelen (1956) are the numbers of millions of revolutions to failure for each of 23 ball bearings. The original data have been put in numerical order for convenience.

17.88	28.92	33.00	41.52	42.12	45.60	48.40	51.84
51.96	54.12	55.56	67.80	68.64	68.64	68.88	84.12
93.12	98.64	105.12	105.84	127.92	128.04	173.40	

The data are clearly positively skewed. The sample mean and standard deviation $\bar{t}$ and s_t are given by

$$\bar{t} = \sum_{i=1}^{23} t_i/23 = 72.22$$

and

$$s_t = \left\{ \sum_{i=1}^{23} (t_i - \bar{t})^2/22 \right\}^{1/2} = 37.49.$$

It is clear from this that an exponential model is inappropriate. From section 2.3 recall that the mean and standard deviation of an exponential random variable are equal. Thus, if an exponential model were appropriate we should expect the corresponding sample-values to be approximately equal. That is,

$$\frac{\bar{t}}{s_t} \cong 1.$$

Here, however, the ratio is nearly 2.

Suppose now we wish to try fitting two slightly less simple models: the Weibull and the lognormal. In both cases it is natural to work with the log-lifetimes. Let $x_i = \log t_i$. Then the sample mean and sample standard deviation of the log-lifetimes are

$$\bar{x} = 4.150; \qquad s_x = 0.534.$$

Recall that, from section 2.4, the mean and standard deviation for log-lifetimes for a Gumbel model are $\mu - \gamma\sigma$ and $\pi\sigma/\sqrt{6}$ respectively where $\gamma = 0.5772\ldots$ is Euler's constant. Thus, a simple estimate for σ, based on the sample moments, is

$$\tilde{\sigma} = \frac{s_x\sqrt{6}}{\pi} = 0.416.$$

An estimate for μ is then

$$\tilde{\mu} = \bar{x} + \gamma\tilde{\sigma} = 4.390.$$

These correspond to estimates for the Weibull parameters:

$$\tilde{\alpha} = \exp(\tilde{\mu}) = 80.64$$

$$\tilde{\eta} = \frac{1}{\tilde{\sigma}} = 2.40.$$

Note here that, had an exponential model been appropriate, we would have expected $\tilde{\eta} \cong 1$.

Alternatively, for a lognormal model, the log-lifetimes are Normally distributed. Hence the sample mean and sample standard deviation of the log-lifetimes are obvious estimates of μ and σ. Thus for a lognormal model we can estimate μ and σ by $\tilde{\mu} = 4.150$ and $\tilde{\sigma} = 0.534$ respectively.

As we shall see shortly, it turns out that the Weibull and lognormal distributions provide satisfactory fits to these data.

Rather than work with sample moments, a different approach is to estimate the survivor function. The **empirical survivor function** $\hat{S}(t)$ is defined by

$$\hat{S}(t) = \frac{\text{\# observations greater than or equal to } t}{n} \qquad (2.13)$$

This is a non-increasing step function with steps at the observed lifetimes. It is a non-parametric estimator of $S(t)$ in the sense that it does not depend on assumptions relating to any specific probability model.

If the value of the survivor function at a particular specified value, t^* say, is of interest, then, using the results for the Binomial distribution in section 1.5, the standard error of $\hat{S}(t^*)$ is

$$\text{se}\{\hat{S}(t^*)\} = \left\{ \frac{\hat{S}(t^*)[1 - \hat{S}(t^*)]}{n} \right\}^{1/2}. \qquad (2.14)$$

Hence an approximate $100(1 - \alpha)\%$ confidence interval for $S(t^*)$ is

$$\hat{S}(t^*) \pm z_{\alpha/2} \, \text{se}\{\hat{S}(t^*)\} \qquad (2.15)$$

where $z_{\alpha/2}$ is the upper $100\alpha/2$ percentage point of the standard Normal distribution.

Furthermore, we can use equations (2.1) and (2.13) to obtain the **empirical cumulative hazard function**

$$\hat{H}(t) = -\log \hat{S}(t).$$

The approximate standard error of $\hat{H}(t^*)$ at a specified value t^* is

$$\text{se}\{\hat{H}(t^*)\} = \left\{ \frac{1 - \hat{S}(t^*)}{n\hat{S}(t^*)} \right\}^{1/2} \qquad (2.16)$$

This expression was obtained from equation (2.14) by using an approach known as the delta method. In order to avoid getting bogged down in generalities at this stage, we defer discussion of the delta method for the moment. Some relevant results on the method are given in section 3.2, and a more rigorous discussion appears in the Appendix, at the end of this book.

It is natural to use $\hat{S}(t)$ and $\hat{H}(t)$ as the basis for graphical displays. First, they stand as descriptions of the data in their own right. Secondly, they may suggest the form of a probability model for the data. Thirdly, after fitting a probability model, they may allow a simple assessment of the adequacy of the fitted model. In addition, in certain circumstances crude parameter estimates for a model may be obtained graphically.

We concentrate for the moment on plots based on $\hat{S}(t)$. One approach is to plot $\hat{S}(t)$ versus t for all t in a convenient range. This gives rise to a rather jagged plot because $\hat{S}(t)$ is a step function with jumps at the observed lifetimes. An alternative formulation is as follows. Suppose the lifetimes have been put in ascending order: $t_{(1)} < t_{(2)} < \cdots < t_{(n)}$ and that there are no tied observations. Then

$$\hat{S}(t_{(i)}) = 1 - \frac{(i-1)}{n}$$

and

$$\hat{S}(t_{(i)} + 0) = 1 - \frac{(i-1)}{n} - \frac{1}{n}.$$

Here $\hat{S}(t_{(i)} + 0)$ means $\hat{S}$ evaluated at a point just greater than $t_{(i)}$. The 'average' value of $\hat{S}$ evaluated near to $t_{(i)}$ is thus

$$\tfrac{1}{2}\{\hat{S}(t_{(i)}) + \hat{S}(t_{(i)} + 0)\} = 1 - \frac{(i-1)}{n} - \frac{1}{2n}$$

$$= 1 - \frac{(i - \tfrac{1}{2})}{n}. \tag{2.17}$$

A commonly used graphical representation of the empirical survivor function, which takes the form of a scatter diagram, is to plot the points $(t_{(i)}, 1 - p_i)$, where the p_i are known as the plotting positions. Various forms for p_i are used in practice, such as $(i-1)/n$, $i/(n+1)$, but, following (2.17) we shall use

$$p_i = \frac{i - \tfrac{1}{2}}{n}.$$

Thus, we shall represent the empirical survivor function graphically by a plot of the points

$$\left(t_{(i)}, 1 - \frac{(i - \tfrac{1}{2})}{n} \right).$$

The case in which there are ties in the data will be subsumed in the general discussion in section 2.11.

For some simple models a transformation of the empirical survivor function plot can be obtained easily so that the transformed plot should be roughly linear if the model is appropriate. Moreover, crude parameter estimates may be obtained from the plot. This approach is particularly useful in the case of Weibull and lognormal models for lifetimes.

Consider first the Weibull model. Then, from (2.3) we have

$$\log\{-\log S(t)\} = \eta \log t - \eta \log \alpha. \tag{2.18}$$

Consequently, if a Weibull model is appropriate for the data, a plot of $\log\{-\log[1-(i-\frac{1}{2})/n]\}$ against $\log t_{(i)}$ should be roughly linear. Further, if the slope and intercept of the plot are a and b respectively, then rough estimates of η and α are a and $\exp(-b/a)$ respectively.

Similarly, for the lognormal model, the survivor function is

$$S(t) = 1 - \Phi\left(\frac{\log t - \mu}{\sigma}\right)$$

where Φ is the standard Normal distribution function. Hence

$$\Phi^{-1}\{1 - S(t)\} = \frac{\log t}{\sigma} - \frac{\mu}{\sigma}. \tag{2.19}$$

So a plot of $\Phi^{-1}\{(i-\frac{1}{2})/n\}$ against $\log t_{(i)}$ should be roughly linear if the lognormal model is appropriate. Again, if the slope and the intercept of the plot are a and b respectively, the rough estimates of μ and σ are $-b/a$ and $1/a$ respectively. Most standard statistical packages allow straightforward calculation of $\Phi^{-1}(x)$. A Fortran program to calculate $\Phi^{-1}(x)$ is given in Beasley and Springer (1977) and reproduced in Griffiths and Hill (1985, p. 191). Alternatively, subroutine libraries such as NAG can be used.

For both the above plots, we have been able to use the fact that the inverse function S^{-1} is readily available without explicitly fitting the model in question. This is not so in general. For example, for a gamma distribution S^{-1} depends crucially on the shape parameter ρ, making these simple graphical methods inappropriate.

An appropriate graphical representation of the empirical cumulative hazard function is to plot $[t_{(i)}, -\log\{1-(i-\frac{1}{2})/n\}]$. This is precisely what was done in Figure 1.1 in the discussions of failure rate in section 1.3. A roughly linear plot through the origin indicates a constant hazard and hence an exponential model since, from (2.2)

$$h(t) = \lambda$$

and hence

$$H(t) = \lambda t.$$

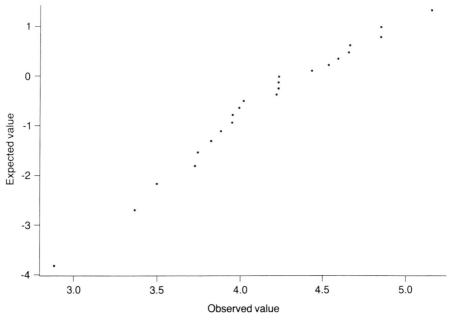

Figure 2.13 Weibull plot for the ball bearings data.

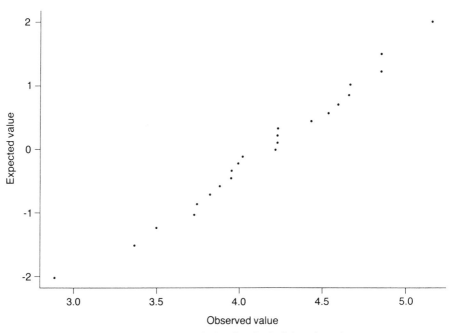

Figure 2.14 Lognormal plot for the ball bearings data.

Example 2.1 (continued)

Figures 2.13 and 2.14 show the plots for the ball bearings data based on (2.18) and (2.19) respectively. The Weibull-based plot, Figure 2.13, certainly does not have unit slope, confirming that an exponential model is inappropriate. Of the two plots, Figure 2.14, based on the lognormal assumption, looks slighty the more linear. This gives an indication that the lognormal model is perhaps preferable to the Weibull model (at least empirically).

Fitting a straight line by eye to Figure 2.13 gives slope 2.3 and intercept -10.0. Thus, rough estimates of η and α are 2.3 and $\exp(10.0/2.3) = 77.3$ respectively. These are in broad agreement with the moment-based estimates of 2.4 and 80.6. Similarly, fitting a straight line by eye to Figure 2.14 gives slope 1.8 and intercept -7.5. Thus, rough estimates for μ and σ in the lognormal model are $7.5/1.8 = 4.2$ and $1/1.8 = 0.56$. Again, these are in broad agreement with the moment-based estimates of 4.15 and 0.53.

2.10 DATA ANALYTIC METHODS: TYPE II CENSORING

Consider the situation in which n units are put on test and observation continues until r units have failed. In other words we have Type II censoring: the first r lifetimes $t_{(1)} < t_{(2)} < \cdots < t_{(r)}$ are observed, but it is known only that the remaining $n-r$ lifetimes exceed $t_{(r)}$. Because $n-r$ observations are 'incomplete', it is impossible to calculate the sample moments. So standard moment-based methods cannot be used. However, all the results based on the empirical survivor function as discussed in section 2.9 still hold. The only difference is that $\hat{S}(t)$ is not defined for $t > t_{(r)}$.

Example 2.2

Mann and Fertig (1973) give details of a life test done on thirteen aircraft components subject to Type I censoring after the tenth failure. The ordered data in hours to failure time are:

$$
\begin{array}{ccccc}
0.22 & 0.50 & 0.88 & 1.00 & 1.32 \\
1.33 & 1.54 & 1.76 & 2.50 & 3.00
\end{array}
$$

Figures 2.15 and 2.16 show the plots of $\log\{-\log[1-(i-\tfrac{1}{2})/n]\}$ and $\Phi^{-1}\{(i-\tfrac{1}{2})/n\}$ respectively against $\log t_{(i)}$. Both plots look reasonably linear, suggesting that it may be rather hard to choose between a Weibull and a lognormal model for these data. This is perhaps not surprising with such a small data set. In addition the slope of the Weibull-based plot looks reasonably close to unity, indicating the plausibility of an exponential model.

In fact, fitting straight lines by eye to Figures 2.15 and 2.16 yields slopes and intercepts 1.4 and -1.1 for the Weibull plot, and 1.0 and -0.5 for the lognormal plot. Hence rough estimates for η and α are 1.4 and $\exp(1.1/1.4) =$

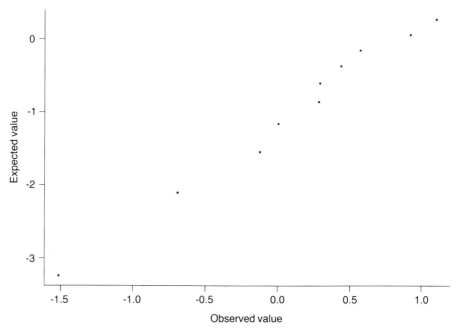

Figure 2.15 Weibull plot for the aircraft components data.

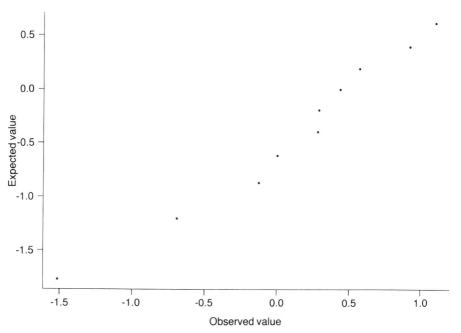

Figure 2.16 Lognormal plot for the aircraft components data.

2.2. In the lognormal model, rough estimates of μ and σ are 0.5 and 1.0 respectively.

Note that simple Type I censoring can also be handled similarly. That is, we observe n units on a life test until a time c has elapsed. If the ordered observed lifetimes are $t_{(1)} < t_{(2)} < \cdots < t_{(r)}$, then it is known only that the remaining $n - r$ lifetimes each exceed c. So $\hat{S}(t)$ is undefined for $t > c$.

2.11 DATA ANALYTIC METHODS: GENERAL CENSORING

We now consider an approach which allows a non-parametric estimate of the survivor function to be obtained in the case of any right-censored sample. The estimate is known as the product-limit (PL) or Kaplan-Meier estimate, after Kaplan and Meier (1958). The results for $\hat{S}(t)$ in sections 2.9 and 2.10 are special cases of the PL estimate.

Suppose there are k distinct times $a_1 < a_2 < \cdots < a_k$ at which failures occur. Let d_j be the number of failures at time a_j. Let n_j be the number of items at risk at a_j; that is, the number of unfailed and uncensored observations just prior to a_j. In additon, let $a_0 = -\infty$, $d_0 = 0$ and $n_0 = n$. Then the PL estimate of $S(t)$ is

$$\hat{S}(t) = \prod{}^{(t)} \left(1 - \frac{d_j}{n_j} \right) \qquad (2.20)$$

where $\prod^{(t)}$ denotes the product over all j such that $a_j < t$. With no censoring present (2.20) reduces to (2.13).

An estimate, based on asymptotic theory for the standard error of $\hat{S}(t^*)$ at a fixed value t^* is

$$\text{se}\{\hat{S}(t^*)\} = \hat{S}(t^*) \left\{ \sum{}^{(t)} \frac{d_i}{n_j(n_j - d_j)} \right\}^{1/2}, \qquad (2.21)$$

where $\sum^{(t)}$ denotes the sum over all j such that $a_j < t$. This is sometimes known as Greenwood's formula; see Greenwood (1926). Approximate confidence limits for $S(t^*)$ may be obtained as in (2.15).

In order to estimate $H(t)$ we can use (2.20) to obtain

$$\hat{H}(t) = -\log \hat{S}(t).$$

Alternatively, a slighty simpler estmate of $H(t)$ is

$$\tilde{H}(t) = \sum{}^{(t)} \frac{d_j}{n_j}.$$

In practice $\hat{H}(t)$ and $\tilde{H}(t)$ are usually close.

At a specified value t^* the standard error of $\hat{H}(t^*)$ and $\tilde{H}(t^*)$ may be approximated, using (2.21) and the delta method, by

$$\text{se}\{\hat{H}(t^*)\} = \text{se}\{\tilde{H}(t^*)\} = \left\{ \sum^{(t)} \frac{d_j}{n_j(n_j - d_j)} \right\}^{1/2}.$$

As in sections 2.9 and 2.10 a plot of $\hat{S}(t)$ versus t can be informative. This may be constructed for all t values, giving a step function, or by plotting only the points $(a_j, 1 - p_j)$ for $j = 1, 2, \ldots, k$ where

$$p_j = 1 - \tfrac{1}{2}\{\hat{S}(a_j) + \hat{S}(a_j + 0)\}.$$

Note that $\hat{S}(a_j + 0) = \hat{S}(a_{j+1})$ for $j = 1, 2, \ldots, k-1$. Similar remarks apply to graphical representation of $\hat{H}(t)$ and $\tilde{H}(t)$.

In the special case of data that are assumed to be Weibull distributed, a plot of the points $(\log a_j, \log\{-\log(1 - p_j)\})$ for $j = 1, 2, \ldots, k$ should be approximately linear if the Weibull model is appropriate. Similarly a plot of the points $(\log a_j, \Phi^{-1}(p_j))$ should be approximately linear if a lognormal model is appropriate. For both plots rough parameter estimates may be obtained as in section 2.9.

Example 2.3

In an experiment to gain information on the strength of a certain type of braided cord after weathering, the strengths of 48 pieces of cord that had been weathered for a specified length of time were investigated. The intention was to obtain the strengths of all 48 pieces of cord. However, seven pieces were damaged during the course of the experiment, thus yielding right-censored strength-values. The strengths of the remaining 41 pieces were satisfactorily observed. Table 2.1 shows the data in coded units from this experiment.

Table 2.1 Strengths in coded units of 48 pieces of weathered braided cord

Uncensored observations									
36.3	41.7	43.9	49.9	50.1	50.8	51.9	52.1	52.3	52.3
52.4	52.6	52.7	53.1	53.6	53.6	53.9	53.9	54.1	54.6
54.8	54.8	55.1	55.4	55.9	56.0	56.1	56.5	56.9	57.1
57.1	57.3	57.7	57.8	58.1	58.9	59.0	59.1	59.6	60.4
60.7									
Right-censored observations									
	26.8	29.6	33.4	35.0	40.0	41.9	42.5		

Two aspects were of particular interest. First, the experimenter felt that it was important that even after weathering the cord-strength should be above 53 in the coded units. Thus, an estimate of $S(53)$ is required. Secondly, previous experience with related strength-testing experiments indicated that a Weibull model might be appropriate here. Thus, a check on the adequacy of the Weibull model is needed, together with estimation of the Weibull parameters if appropriate.

These questions will be addressed further in Chapter 3, but for now we use the PL estimate for the data to obtain $\hat{S}(53)$ and to investigate the Weibull model graphically.

Table 2.2 shows the first few lines of the calculation of $\hat{S}$ and related quantities for illustration. We see that, from equation (2.20) and Table 2.2,

$$\hat{S}(53) = 0.6849.$$

Table 2.2 Sample calculations of $\hat{S}$ and related quantities for Example 2.3

j	a_j	n_j	d_j	$(n_j - d_j)/n_j$	$\hat{S}(a_j + 0)$	$d_j/\{n_j(n_j - d_j)\}$
0	$-\infty$	48	0	1.0000	1.0000	0.0000
1	36.3	44	1	0.9773	0.9773	0.0005
2	41.7	42	1	0.9762	0.9540	0.0006
3	43.9	39	1	0.9744	0.9295	0.0007
4	49.9	38	1	0.9737	0.9051	0.0007
5	50.1	37	1	0.9730	0.8806	0.0008
6	50.8	36	1	0.9722	0.8562	0.0008
7	51.9	35	1	0.9714	0.8317	0.0008
8	52.1	34	1	0.9706	0.8072	0.0009
9	52.3	33	2	0.9394	0.7583	0.0020
10	52.4	31	1	0.9677	0.7338	0.0011
11	52.6	30	1	0.9667	0.7094	0.0011
12	52.7	29	1	0.9655	0.6849	0.0012
13	53.1	28	1	0.9643	0.6605	0.0013

The approximate standard error of $\hat{S}(53)$ is, from equation (2.21) and Table 2.2,

$$\mathrm{se}\{\hat{S}(53)\} = 0.6849\{0.0112\}^{1/2}$$
$$= 0.0725.$$

Thus an approximate 95% confidence interval for $S(53)$ is

$$0.6849 \pm 1.96 \times 0.0725.$$

That is,

$$(0.54, 0.83).$$

Note that the four smallest censored values have been in effect ignored in this analysis as they were censored before the first uncensored observation.

Figure 2.17 shows a plot of $(\log a_j, \log\{-\log(1-p_j)\})$. The bulk of this plot seems to be linear. However, the points corresponding to the three lowest strengths lie considerably above the apparent line. Whilst these points have great visual impact because they are somewhat isolated from the rest of the plot, they are also the least reliable since they correspond to the extreme lower tail where the data are somewhat sparse. In this particular example the extreme points are made even less reliable in view of the relatively large amount of censoring at low strength-levels. The overall lack of curvature in the plot apart from the three isolated points suggests that a Weibull model should not be ruled out. If we assume a Weibull model for the moment, fitting a line by eye to the plot (ignoring the isolated points) gives slope 18.5 and intercept -75. Thus, rough estimates of η and α are 18.5 and 57.6 respectively.

Figure 2.18 shows a plot of $(\log a_j, \Phi^{-1}(p_j))$. Here there is some slight indication of curvature in the plot, even after ignoring the three isolated points.

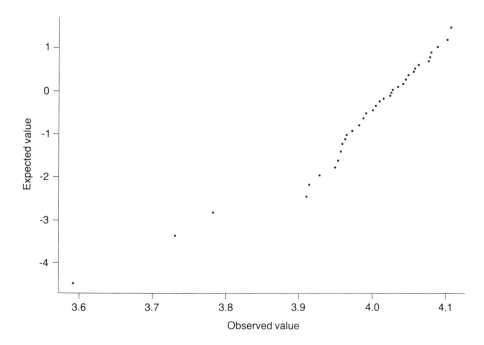

Figure 2.17 Weibull plot for the cord strength data.

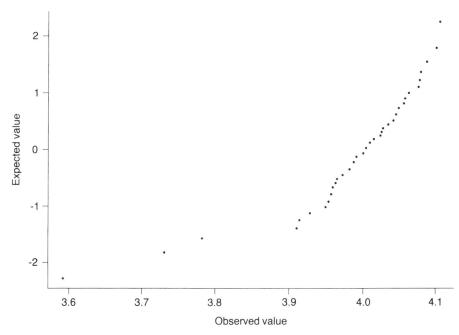

Figure 2.18 Lognormal plot for the cord strength data.

Hence a lognormal model appears to be somewhat less satisfactory than the Weibull model.

In this section we have discussed non-parametric estimation, that is estimation which does not require specification of a particular parametric model, of the survivor function and the cumulative hazard function in the presence of right-censored observations. A related approach for estimating quantiles, such as the median and the quartiles, in the presence of right-censoring is given in Kimber (1990).

When data contain left-censored or interval-censored observations, results analogous to the PL estimator of the survivor function are available. However, they are considerably more complex than the PL estimator and they have not been used much in the reliability context. One reason for this is the relative rarity of the occurrence of left-censored observations in reliability. In addition, unless the intervals are very wide, interval-censoring is generally relatively unimportant in practical terms. In fact all observations of continuous variables, such as time and strength, are interval-censored since data are only recorded to a finite number of decimal places. However, this aspect is usually ignored in analyses. For further information on non-parametric estimation of the survivor function in the presence of left-censored or interval-censored observations the reader is referred to Turnbull (1974, 1976).

3
Statistical methods for single samples

3.1 INTRODUCTION

Towards the end of Chapter 2 some simple statistical methods were introduced. These methods are such that they may be used before embarking on a more formal statistical analysis. In this chapter we first discuss methods for obtaining statistical inferences in the reliability context. In section 3.2 the method of maximum likelihood estimation is discussed in general terms. Some particular illustrations are given in section 3.3. Likelihood-based methods for hypothesis testing and confidence regions are then introduced in section 3.4. We then make some general remarks on likelihood-based methods in section 3.5. Finally we discuss in section 3.6 some methods that may be applied after fitting a parametric model, such as a Weibull distribution, in order to assess the adequacy of the fitted model.

3.2 MAXIMUM LIKELIHOOD ESTIMATION: GENERALITIES

In this section we give details of a general method of estimation of parameters, called maximum likelihood estimation (ML). To fix ideas suppose we have a sample of observations $t_1, t_2, \ldots, t_n$ from the population of interest. For the moment we assume that none of the observations is censored. In the reliability context it is reasonable to assume that the t_i are lifetimes. Suppose also that they can be regarded as observations with common density function $f(t; \theta_1, \theta_2, \ldots, \theta_m)$ where the form of f is known but where the parameters $\theta_1, \theta_2, \ldots, \theta_m$ are unknown. So, for example, we may perhaps assume that the observations are Weibull-distributed with unknown η and α. For brevity we shall denote dependence on $\theta_1, \theta_2, \ldots, \theta_m$ by θ, so that the common density may be written $f(t; \theta)$. Then the likelihood of the observations is defined by

$$L(\theta) = \prod_{i=1}^{n} f(t_i; \theta).$$

More generally, suppose that some of the observations are right-censored. Then we can split the observation numbers $1, 2, \ldots, n$ into two disjoint sets, one, U say, corresponding to observations that are uncensored, the other, C say, corresponding to right-censored observations. Then the likelihood in this case is defined by

$$L(\theta) = \left\{ \prod_{i \in U} f(t_i; \theta) \right\} \left\{ \prod_{i \in C} S(t_i; \theta) \right\}. \tag{3.1}$$

Thus, for a right-censored observation the density has been replaced by the survivor function. In a similar manner for a left-censored observation the density should be replaced by the distribution function. For an interval-censored observation the density should be replaced by the distribution function evaluated at the upper end-point of the interval minus the distribution function evaluated at the lower end-point of the interval, thus yielding the probability of occurrence of a lifetime within the interval.

It is almost always more convenient to work with the log-likelihood, $l(\theta)$ defined by

$$l(\theta) = \log L(\theta).$$

The maximum likelihood estimates (MLEs) $\hat{\theta}_1, \hat{\theta}_2, \ldots, \hat{\theta}_m$ of $\theta_1, \theta_2, \ldots, \theta_m$ are those values that maximize the likelihood, or equivalently, the log-likelihood. Alternatively, and more usually, the MLEs may be found by solving the likelihood equations

$$\frac{\partial l}{\partial \theta_j} = 0 \quad (j = 1, 2, \ldots, m).$$

Both approaches will usually involve numerical methods such as Newton or quasi-Newton algorithms. For most of the problems covered in this book the necessary numerical methods will be available in a package such as GLIM (see also Aitkin and Clayton, 1980), in subroutine libraries such as NAG and IMSL, and in the literature; see Press *et al.* (1986). In most simple situations (e.g. fitting a two parameter Weibull distribution) direct maximization of L or l will yield identical results to solving the likelihood equations. However, there exist situations where one or other of the two methods is unsatisfactory; see section 3.5.

Suppose that $\hat{\theta} = (\hat{\theta}_1, \hat{\theta}_2, \ldots, \hat{\theta}_m)$ has been calculated. We may be interested in some function of the unknown parameters such as

$$\phi = g(\theta)$$

where g is a specified one-to-one function. Then the MLE of ϕ is $\hat{\phi}$ defined by

$$\hat{\phi} = g(\hat{\theta}).$$

For example, one is often interested in estimating a quantile of the lifetime distribution; that is, estimating

$$q(p) \equiv q(p; \theta),$$

satisfying

$$\Pr\{T \geq q(p)\} = S\{q(p)\} = p,$$

where $0 < p < 1$ is specified. Hence the MLE of the quantile is just $q(p; \hat{\theta})$.

Furthermore, from asymptotic theory the precision of the MLEs may in many cases be estimated in a routine way. Consider the $m \times m$ observed information matrix $\mathbf{J}$ with entries

$$\frac{-\partial^2 l}{\partial \theta_j \partial \theta_k} \quad (j = 1, 2, \ldots, m;\ k = 1, 2, \ldots, m) \tag{3.2}$$

evaluated at $\hat{\theta}$. Then the inverse of $\mathbf{J}$ is the estimated variance-covariance matrix of $\hat{\theta}_1, \hat{\theta}_2, \ldots, \hat{\theta}_m$. That is, if $\mathbf{V} = \mathbf{J}^{-1}$ has entries v_{jk}, then v_{jk} is the estimated covariance between $\hat{\theta}_j$ and $\hat{\theta}_k$. In particular an estimate for the standard error of $\hat{\theta}_j$ $(j = 1, 2, \ldots, m)$ is just $v_{jj}^{1/2}$.

In addition, if $\phi = g(\theta)$, then the standard error of $\hat{\phi}$ may be estimated by

$$\text{se}(\hat{\phi}) = \left\{ \sum_{j=1}^{m} \sum_{k=1}^{m} (\partial g / \partial \theta_j)(\partial g / \partial \theta_k) v_{jk} \right\}^{1/2}, \tag{3.3}$$

where the partial derivatives are evaluated at $\hat{\theta}$. This procedure is often referred to as the **delta method**. In the special case where $m = 1$, so that θ is a scalar parameter, equation (3.3) reduces to

$$\text{se}(\hat{\phi}) \cong \left| \frac{\mathrm{d}g}{\mathrm{d}\theta} \right| \sqrt{v_{11}}, \tag{3.4}$$

where $\mathrm{d}g/\mathrm{d}\theta$ is evaluated at $\hat{\theta}$. It is this equation (3.4) that was used in section 2.9 to obtain equation (2.16) from equation (2.14) with $g(\theta) = -\log \theta$. A more detailed discussion of the delta method is given in the Appendix at the end of this book.

Whilst construction of standard errors on the basis of equations (3.2), (3.3) and (3.4) is usually straightforward, the method does have certain drawbacks. These are discussed more fully in section 3.4.

Of course, ML is not the only estimation method available. We have already seen some ad hoc methods in sections 2.9 to 2.11. However, from the point of view of the user, ML has several major advantages. First its generality ensures that most statistical problems of estimation likely to arise in the reliability context may be dealt with using ML. Many other methods, such as those based on linear functions of order statistics (see David, 1981), are very simple to use in some univariate problems but are difficult or impossible to generalize to more complex situations. In addition, the generality of ML is an advantage from a computational point of view since, if desired, essentially the same program may be used to obtain MLEs whatever the context. Secondly, the functional invariance property of MLEs ensures that, having calculated $\hat{\theta}$, one may obtain the MLE of $g(\theta)$ immediately without having to restart the estimation process. Thirdly, approximate standard errors of MLEs may be found routinely by inversion of the observed information matrix.

From the theoretical point of view ML also has some properties to recommend it. Under mild regularity conditions MLEs are consistent, asymptotically Normal and asymptotically efficient. Roughly speaking these results mean that if the whole population is observed ML will give exactly the right answer, and that in large samples a MLE is approximately Normally distributed, approximately unbiased and with the smallest attainable variance. For technical details the reader should consult Cox and Hinkley (1974).

3.3 MAXIMUM LIKELIHOOD ESTIMATION: ILLUSTRATIONS

In this section we illustrate the calculation of MLEs in certain special cases. Throughout we shall assume that we have a single sample of observations, possibly right-censored, and that these observations are identically distributed. The case in which the parameters of interest depend on some explanatory or regressor variables will be covered in Chapters 4 and 5.

We assume that in the sample of possibly right-censored lifetimes $t_1, t_2, \ldots, t_n$, there are r uncensored observations and $n-r$ right-censored observations. We also define $x_i = \log t_i$ $(i = 1, 2, \ldots, n)$.

Exponential distribution
The log-likelihood is, from equations (2.2) and (3.1),

$$l(\lambda) = r \log \lambda - \lambda \sum_{i=1}^{n} t_i.$$

Hence

$$\frac{dl}{d\lambda} = \frac{r}{\lambda} - \sum_{i=1}^{n} t_i.$$

This may be set to zero and solved immediately to give

$$\hat{\lambda} = \frac{r}{\sum\limits_{i=1}^{n} t_i}. \tag{3.5}$$

Notice that the denominator is the total time on test(TTT). Also,

$$\frac{-d^2l}{d\lambda^2} = \frac{r}{\lambda^2}. \tag{3.6}$$

Hence the estimated standard error of $\hat{\lambda}$ is $\hat{\lambda}/\sqrt{r}$. Notice that it is necessary that $r>0$, that is, at least one lifetime must be uncensored.

As special cases, if all the observations are uncensored then λ is just the reciprocal of the sample mean, whereas if only the r smallest lifetimes $t_{(1)}<t_{(2)}<\cdots<t_{(r)}$ have been observed (simple Type II censoring) then

$$\hat{\lambda} = r \bigg/ \left\{ \sum_{i=1}^{r} t_{(i)} + (n-r)t_{(r)} \right\}. \tag{3.7}$$

The reciprocal of the right-hand side in (3.7) is sometimes known as the one-sided Winsorized mean.

Weibull distribution
The log-likelihood for a Weibull sample is, from equations (2.3), (2.4) and (3.1)

$$l(\eta, \alpha) = r \log \eta - r\eta \log \alpha + (\eta - 1) \sum_{u} \log t_i - \alpha^{-\eta} \sum_{i=1}^{n} t_i^{\eta}.$$

Alternatively, letting $x_i = \log t_i$ and using a Gumbel formulation of the problem, we obtain from equations (2.6), (2.7) and (3.1)

$$l(\mu, \sigma) = -r \log \sigma + \sum_{u} (x_i/\sigma) - (r\mu/\sigma) - \sum_{i=1}^{n} \exp\{(x_i - \mu)/\sigma\}.$$

Thus

$$\sigma\frac{\partial l}{\partial \mu} = -r + \sum_{i=1}^{n} \exp\{(x_i - \mu)/\sigma\}$$

$$\sigma^2\frac{\partial l}{\partial \sigma} = -r\sigma - \sum_{u} x_i + r\mu + \sum_{i=1}^{n} \exp\{(x_i - \mu)/\sigma\}(x_i - \mu).$$

Hence

$$\hat{\mu} = \hat{\sigma}\log\left\{\frac{1}{r}\sum_{i=1}^{n}\exp(x_i/\hat{\sigma})\right\} \tag{3.8}$$

and

$$\frac{1}{r}\sum_{u} x_i + \hat{\sigma} - \sum_{i=1}^{n} x_i\exp(x_i/\hat{\sigma})\bigg/\sum_{i=1}^{n}\exp(x_i/\hat{\sigma}) = 0 \tag{3.9}$$

Note that equation (3.9) does not involve $\hat{\mu}$. So the problem of obtaining $\hat{\mu}$ and $\hat{\sigma}$ reduces simply to finding $\hat{\sigma}$, after which $\hat{\mu}$ may be found directly from equation (3.8). The solution to equation (3.9) must be found numerically. Routines for solving such non-linear equations are readily available in subroutine libraries such as NAG and IMSL. See also Press *et al.* (1986). A further possibility is to find two values of σ by trial and error which give opposite signs to the left side of equation (3.9). These may be used as starting values in a repeated bisection scheme. This can easily be programmed on even a very small computer.

The second derivatives of l are

$$\frac{-\partial^2 l}{\partial\mu^2} = \frac{1}{\sigma^2}\exp(-\mu/\sigma)\sum_{i=1}^{n}\exp(x_i/\sigma)$$

$$\frac{-\partial^2 l}{\partial\mu\partial\sigma} = \frac{-r}{\sigma^2} + \frac{1}{\sigma^3}\exp(-\mu/\sigma)(\sigma-\mu)\sum_{i=1}^{n}\exp(x_i/\sigma) + \frac{1}{\sigma^3}\exp(-\mu/\sigma)\sum_{i=1}^{n}x_i\exp(x_i/\sigma)$$

$$\frac{-\partial^2 l}{\partial\sigma^2} = \frac{-r}{\sigma^2} - 2\sum_{u}\frac{x_i}{\sigma^3} + \frac{2r\mu}{\sigma^3} + \frac{1}{\sigma^4}\sum_{i=1}^{n}\exp(x_i/\sigma)(x_i-\mu)\{2\sigma + x_i - \mu\}.$$

These expressions simplify considerably when they are evaluated at $(\mu, \sigma) =$

$(\hat{\mu}, \hat{\sigma})$ to give

$$\frac{-\partial^2 l}{\partial \mu^2} = \frac{r}{\hat{\sigma}^2}$$

$$\frac{-\partial^2 l}{\partial \mu \partial \sigma} = \sum_{i=1}^{n} \left(\frac{x_i - \hat{\mu}}{\hat{\sigma}}\right) \exp\left\{\left(\frac{x_i - \hat{\mu}}{\hat{\sigma}}\right)\right\}$$

$$\frac{-\partial^2 l}{\partial \sigma^2} = r + \sum_{i=1}^{n} \left(\frac{x_i - \hat{\mu}}{\hat{\sigma}}\right)^2 \exp\left\{\left(\frac{x_i - \hat{\mu}}{\hat{\sigma}}\right)\right\}.$$

Example 3.1

Consider again the data first encountered in Example 2.2. For illustration purposes we now fit an exponential model using ML. Here $n = 13$, $r = 10$ and

$$\sum_{i=1}^{n} t_i = 23.05$$

Hence, using equations (3.5) and (3.6) for the exponential model the MLE for λ is $\hat{\lambda} = 10/23.05 = 0.434$, with standard error $\hat{\lambda}/\sqrt{r} = 0.137$. Figure 3.1 shows a plot of the log-likelihood as a function of λ. Note that there is a single maximum and that the function is rather skewed.

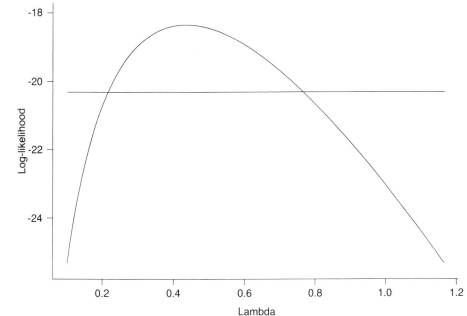

Figure 3.1 Log-likelihood for the aircraft components data with an exponential model. The line $l(\hat{\lambda})$-1.92 has been marked to show the 95 per cent confidence interval for λ based on W.

Example 3.2

Consider again the data first encountered in Example 2.1. We shall initially be concerned with fitting a parametric model to these data using for illustration first the Weibull distribution and then the lognormal distribution. For convenience we shall work with the logged data which amounts to fitting a Gumbel distribution and a Normal distribution.

For the Gumbel distribution we used a simple iterative scheme via the NAG library to obtain $\hat{\mu} = 4.405$, $\hat{\sigma} = 0.476$ with estimated variance-covariance matrix

$$\mathbf{V}_{\text{Gumbel}} = \begin{pmatrix} 0.01104 & -0.00257 \\ -0.00257 & 0.00554 \end{pmatrix}.$$

For the Normal distribution $\hat{\mu}$ and $\hat{\sigma}$ are respectively the sample mean and sample standard deviation (with divisor n rather than $n-1$), giving $\hat{\mu} = 4.150$ and $\hat{\sigma} = 0.522$ and

$$\mathbf{V}_{\text{Normal}} = \begin{pmatrix} 0.01184 & 0 \\ 0 & 0.00592 \end{pmatrix}.$$

Note that the parameters μ and σ are used here as generic notations for location and scale parameters for log-lifetimes. There is no reason why, say, μ in the Gumbel formulation should be equal to μ in the Normal formulation.

Suppose that various quantiles are of interest: the median, the lower 10% point and the lower 1% point. For both the distributions fitted to the log-lifetimes a quantile $q(p)$ is of the form $\mu + \sigma a(p)$, where $a(p)$ is readily available in standard statistical tables in the Normal case and where $a(p) = \log(-\log p)$ in the Gumbel case. Table 3.1 gives the numerical values of $a(p)$ for these two distributions for the required values of p, namely 0.5, 0.9 and 0.99. Using the given values of $a(p)$ and the calculated MLEs we can obtain the estimated quantiles for the log-lifetimes.

Table 3.1 Values of $a(p)$ in the Normal and Gumbel cases with $p=0.5$, 0.9, 0.99

p	Gumbel $a(p)$	Normal $a(p)$
0.5	−0.367	0
0.9	−2.250	−1.282
0.99	−4.600	−2.326

These may be transformed back to the original scale of measurement by exponentiating. For example, the estimated median of the log-lifetimes in the Gumbel case is given by $4.405 - 0.367 \times 0.476 = 4.230$. Thus, in the Gumbel case the estimated median lifetime is exp(4.320) = 68.7 million revolutions. Table 3.2 shows the estimated quantiles for the lifetimes using the two parametric formulations, together with standard errors using equation (3.3) with $g(\mu, \sigma) = \exp\{\mu + \sigma a(p)\}$.

Table 3.2 Quantile estimates (in millions of revolutions) for Weibull and lognormal models, together with their standard errors

Quantile	Weibull estimate	Lognormal estimate
Median	68.7 (8.0)	63.4 (6.9)
Lower 10%	28.1 (6.3)	32.5 (4.8)
Lower 1%	9.2 (3.6)	18.1 (3.9)

Examination of Table 3.2 shows that, especially in the case of the most extreme quantile, the two parametric models appear to give very different estimates. This situation is not unusual.

3.4 TESTS AND CONFIDENCE REGIONS BASED ON LIKELIHOOD

In Example 3.2 above we have seen that fitting different parametric models may give very different estimates of a quantity of interest, such as a quantile. Thus, some methods for choosing between parametric models are clearly required. One approach is based on asymptotic properties of the likelihood function. These properties also enable confidence intervals, or more generally, confidence regions, to be calculated. First, we shall state the main general results. Then we shall give some simple illustrations using the data already discussed above.

We begin by supposing that the parametric model of interest depends on parameters $\theta_1, \theta_2, ..., \theta_m$. Suppose we are interested in testing or making confidence statements about a subset $\theta^{(A)}$ of these parameters. Label the remaining parameters $\theta^{(B)}$. Of course, $\theta^{(A)}$ may contain all of the m parameters, so that $\theta^{(B)}$ is empty. Let $(\hat{\theta}^{(A)}, \hat{\theta}^{(B)})$ be the joint MLE of $(\theta^{(A)}, \theta^{(B)})$. Let $\hat{\theta}^{(B)}(A_0)$ be the MLE of $\theta^{(B)}$ when $\theta^{(A)}$ is fixed at some chosen value $\theta_0^{(A)}$, say. Then two likelihood-based methods for testing and constructing confidence regions are as follows:

1. Let

$$W(\theta_0^{(A)}) = W = 2\{l(\hat{\theta}^{(A)}, \hat{\theta}^{(B)}) - l[\theta_0^{(A)}, \hat{\theta}^{(B)}(A_0)]\}.$$

Then under the null hypothesis $\theta^{(A)} = \theta_0^{(A)}$, W has approximately a chi-squared distribution with m_a degrees of freedom, where m_a is the dimension of $\theta^{(A)}$. Large values of W relative to $\chi^2(m_a)$ supply evidence against the null hypothesis. The corresponding $1 - \alpha$ confidence region for $\theta^{(A)}$ is

$$\{\theta^{(A)}: W(\theta^{(A)}) \leq \chi_{\bar{\alpha}}^2(m_a)\},$$

where $\chi_{\bar{\alpha}}^2(m_a)$ is the upper 100α percentage point of $\chi^2(m_a)$.

2. Suppose $\mathbf{V} = \mathbf{V}(\hat{\theta}^{(A)}, \hat{\theta}^{(B)})$ is the variance-covariance matrix for $(\hat{\theta}^{(A)}, \hat{\theta}^{(B)})$ evaluated at the MLE, as in section 3.3. Let $\mathbf{V}_A = \mathbf{V}_A(\hat{\theta}^{(A)}, \hat{\theta}^{(B)})$ be the leading submatrix of $\mathbf{V}$ corresponding to $\hat{\theta}^{(A)}$. That is, $\mathbf{V}_A$ is the submatrix of $\mathbf{V}$ corresponding to the estimated variance and covariance of $\hat{\theta}^{(A)}$. Then

$$W^*(\theta_0^{(A)}) = (\hat{\theta}^{(A)} - \theta_0^{(A)})^T \mathbf{V}_A^{-1} (\hat{\theta}^{(A)} - \theta_0^{(A)})$$

also has an approximate $\chi^2(m_a)$ distribution under the null hypothesis $\theta^{(A)} = \theta_0^{(A)}$. The corresponding approximate $1 - \alpha$ confidence region for $\theta^{(A)}$ is given by

$$\{\theta^{(A)}: W^*(\theta^{(A)}) \leq \chi_{\bar{\alpha}}^2(m_a)\}.$$

In the special case when $\hat{\theta}^{(A)}$ is a scalar this leads to a symmetric $1 - \alpha$ confidence interval

$$\hat{\theta}^{(A)} \pm z_{\alpha/2} V_A^{1/2}$$

where $z_{\alpha/2}$ is the upper $100\alpha/2$ percentage point of the standard Normal distribution.

Both methods based on W and W^* respectively are asymptotically equivalent, and often give very similar results in practice. However, large discrepancies are possible. In such cases the method based on W is preferable because the results are invariant to reparametrization and the shape of the confidence region is essentially decided by the data. Confidence regions based on W^* are necessarily elliptical in the parametrization used but will yield non-elliptical regions under non-linear parameter transformations.

Example 3.1 (continued)
Having fitted an exponential model to the Mann and Fertig data, we shall now calculate a 95% confidence interval for λ. In the notation we used in the general case $\theta^{(A)} = \lambda$, $\theta^{(B)}$ is redundant, and $m_a = 1$. Here

$$l(\lambda) = 10 \log \lambda - 23.05\lambda$$

and

$$l(\hat{\lambda}) = -18.35.$$

Hence

$$W(\lambda) = 2\{-18.35 - 10 \log \lambda + 23.05\lambda\}.$$

Thus, a 95% confidence interval for λ based on W is $\{\lambda: W(\lambda) \leq 3.84\}$ since 3.84 is the upper 5% point of $\chi^2(1)$. In other words the required confidence interval consists of all those values of λ such that $l(\lambda)$ is within 1.92 of $l(\hat{\lambda})$. This interval is marked on Figure 3.1 and corresponds to [0.22, 0.76]. Note that because of the skewness of the log-likelihood, this interval is not centred at the MLE, 0.434.

A second method of calculating a 95% confidence interval for λ is to use W^*. This gives limits $0.434 \pm 1.96 \times 0.137$, where 1.96 is the upper 2.5% point of the standard Normal distribution and 0.137 is the standard error of $\hat{\lambda}$. Thus, the interval based on W^* is [0.17, 0.70]. To show the dependence of W^* on the particular parametrization used, consider first reparametrizing the exponential distribution in terms of $\alpha = 1/\lambda$. Then the MLE for α is $\hat{\alpha} = 1/\hat{\lambda} = 1/0.434 = 2.305$. Further, the standard error of $\hat{\alpha}$ may be found using equation (3.4), giving

$$se(\hat{\alpha}) = \frac{\hat{\alpha}}{\sqrt{r}} = 0.729.$$

Hence, using W^*, a 95% confidence interval for α is $2.305 \pm 1.96 \times 0.729$; that is, [0.88, 3.73]. Hence the corresponding confidence interval for λ is [1/3.73, 1/0.88] i.e. [0.27, 1.14].

Thus we have three quite different confidence intervals for λ. As remarked above, the interval based on W is generally to be preferred. In the special case of the exponential distribution some exact distribution theory for the MLE is available, which gives an exact 95% confidence interval for λ as [0.21, 0.74], which is in good agreement with the w-based interval. The exact interval is based upon the fact that $2r\lambda/\hat{\lambda}$ has a $\chi^2(2r)$ distribution. Note also that, had a higher degree of confidence been required, the W^*-based intervals would have contained negative values, clearly a nonsense since λ is necessarily positive. The occurrence of negative values in a confidence interval for a necessarily positive parameter is not possible when using the W-based method.

Example 3.2 (continued)
Earlier we fitted a Weibull model to the Lieblein and Zelen data. Could we simplify the model and assume exponentiality? Equivalently, working with the log-lifetimes, could we fix $\sigma = 1$ in the Gumbel model? We can take $\theta^{(A)} = \sigma$, $\theta^{(B)} = \mu$

and $m_a = 1$ in the general framework and test the null hypothesis $\sigma = 1$.

We begin by considering the testing procedure based on W^*, results for which are virtually immediate since we have already estimated the variance-covariance matrix of $(\hat{\mu}, \hat{\sigma})$ as $\mathbf{V}_{\text{Gumbel}}$. Here

$$W^*(1) = (0.476 - 1)^2/0.00554 = 49.56.$$

This is highly significant as $\chi^2(1)$, indicating that an exponential model is inappropriate. The result is so clear cut that it is probably not worth using the more accurate test based on w in this particular numerical example. However, we shall do so in the interest of illustrating the method. Let $\hat{\mu}_0$ be the MLE of μ when $\sigma = 1$ is fixed. Then

$$l(\mu, 1) = \sum_u x_i - r\mu - \sum_{i=1}^{n} \exp(x_i - \mu)$$

$$\frac{dl}{d\mu} = -r + \sum_{i=1}^{n} \exp(x_i - \mu).$$

Hence

$$\hat{\mu}_0 = \log\left\{ \frac{1}{r} \sum_{i=1}^{n} \exp(x_i) \right\}.$$

In this example, $\hat{\mu}_0 = 4.280$, so that

$$l(\hat{\mu}_0, 1) = \sum_u x_i - r\hat{\mu}_0 - r = -25.99.$$

Also, $l(\hat{\mu}, \hat{\sigma}) = -18.24$, giving

$$W(1) = 2\{l(\hat{\mu}, \hat{\sigma}) - l(\hat{\mu}_0, 1)\} = 15.50.$$

Again, this is highly significant as $\chi^2(1)$, leading to the same inference as that based on W^*, though the result is not so extreme. These results confirm the earlier informal analysis discussed in Example 2.1.

Example 3.2 (continued) gives an illustration of testing nested hypotheses: the model specified in the null hypothesis is a special case of the model specified in the alternative. In the example the exponential model is a special case of the Weibull model. Other examples include testing:

1. an exponential model against a gamma model;
2. a Weibull model against a generalized gamma model; and
3. an exponential model with specified λ against an exponential model with unspecified λ.

This situation also arises commonly in regression analysis (see Chapters 4, 5, and 6) in which the effect of an explanatory variable may be assessed by testing that its coefficient is zero against the alternative that its coefficient is non-zero.

What of the situation in which the hypotheses to be tested are non-nested? In the context of this chapter, where we are concerned with methods for univariate data, such a situation might arise in attempting to choose between two separate families of distributions. For example, we might wish to choose between a Weibull model and a lognormal model. Unfortunately the distribution theory of section 3.4 no longer holds in general in this situation. Thus, one is faced with the choice of either circumventing the problem or using theory specifically designed for this particular problem. To fix ideas we stay with the choice between Weibull and lognormal models.

An informal approach is to use data analytic methods such as those outlined in sections 2.9 and 2.11. Another approach is to use goodness-of-fit tests for the Weibull distribution and for the lognormal distribution. A major reference in this field is D'Agostino and Stephens (1986). Yet another method is to fit a more comprehensive model which contains the competing models. Possibilities include the generalized gamma model and a mixture model, both discussed in section 2.7. The hypotheses Weibull versus general alternative and lognormal versus general alternative can then be tested. If only one of the competing models can be rejected, then the other one is preferable. However, it is perfectly possible that neither model can be rejected.

On the other hand we can face up squarely to the problem of testing separate families of distributions. First, it must be decided which distribution should be the null hypothesis. For example, if in the past similar data sets have been well fitted by a Weibull model, then it is natural to take a Weibull distribution as the null model. Pioneering work on this topic appears in Cox (1961, 1962b) and Atkinson (1970). In the reliability context, see Dumonceaux and Antle (1973) and Dumonceaux, Antle and Haas (1973). Typically, however, these tests, which are based on maximized likelihoods, have rather low power unless the sample size is large. When there is no natural null hypothesis a pragmatic approach is to test Weibull versus lognormal and lognormal versus Weibull. Once again, it is possible that neither distribution may be rejected in favour of the other.

A different approach is to force a choice between competing models; see Siswadi and Quesenberry (1982). This procedure, however, is not recommended, especially if estimation of extreme quantiles is the aim of the analysis. To force a choice between two or more models, which all fit a set of data about equally well in some overall sense, can lead to spuriously precise inferences about extreme quantiles (see Examples 2.9 and 3.2).

Example 3.2 (continued[2])

We have seen in Example 2.1 that on the basis of simple plots both the Weibull and lognormal models appear to fit the data well. However, in Example 3.2 we have seen that estimates of extreme quantiles for the two models are very different.

The maximized log-likelihoods are, working with the log-lifetimes,

$$\begin{array}{ll} \text{Weibull} & -18.24 \\ \text{lognormal} & -17.27. \end{array}$$

Thus, on the basis of maximized log-likelihood the lognormal model, as with the plots, appears slightly better than the Weibull model. However, using the results of Dumonceaux and Antle (1973), we cannot reject the Weibull model in favour of the lognormal model or vice versa (using a 5% significance level for each test). So we are still unable to choose between the two models on statistical grounds.

If, for example, the median is of major concern, the choice between the models is relatively unimportant. Alternatively, a non-parametric estimate could be used; see sections 2.9 and 2.11 and Kimber (1990). However, if the lower 1% quantile is of prime interest the non-parametric estimate is not an option. If a choice must be made between models, then the more pessimistic one is probably preferable (though this depends on the context of the analysis). However, the real message is that the data alone are insufficient to make satisfactory inferences about low quantiles. If quantifiable information is available in addition to the sample data, then a Bayesian approach (see Chapter 6) may be fruitful. Another way round the problem is to collect more data so that satisfactory inferences about low quantiles may be drawn. The approach of Smith and Weissman (1985) to estimation of the lower tail of a distribution using only the smallest k ordered observations is another possibility, thus side-stepping the problem of a fully parametric analysis.

3.5 REMARKS ON LIKELIHOOD-BASED METHODS

Likelihood-based procedures have been discussed above in relation to parametric analyses of a single sample of data. These results generalize in a natural way for more complex situations, such as when information on explanatory variables is available as well as on lifetimes. These aspects are covered in later chapters.

The asymptotic theory which yields the Normal and χ^2 approximations used above requires certain conditions on the likelihood functions to be satisfied. The situation most relevant to reliability in which the regularity conditions on the likelihood do not hold occurs when a guarantee parameter must be estimated. This can cause some problems, which have been addressed by many authors including Smith and Naylor (1987) for the three-parameter Weibull distribution, Eastham *et al.* (1987) and LaRiccia and Kindermann (1983) for the three-parameter lognormal distribution, and Kappenman (1985) for three-parameter Weibull, lognormal and gamma distributions. Cheng and Amin (1983) also discuss problems with guarantee parameters.

For the relatively simple models discussed so far (e.g. exponential, Weibull, Gumbel, Normal, lognormal) the regularity conditions on the likelihood do hold. For some of these models (e.g. Normal and lognormal with no censoring) closed form MLEs exist. However, in most likelihood-based analyses of reliability data some iterative scheme is needed. Essentially one requires a general program to handle the relevant data, together with a function maximizing procedure, such as quasi-Newton methods in NAG or Press *et al.* (1986). To fit a specific model all that is necessary is to 'bolt on' a subroutine to evaluate the relevant log-likelihood function, and possibly the first and second derivatives of the log-likelihood. For some models some numerical integration or approximation may be needed (e.g. polygamma functions for the gamma distribution, Normal survivor function for the Normal distribution with censored observations). This may be programmed from the relevant mathematical results (Abramowitz and Stegun, 1972) via available algorithms (Press *et al.* 1986; Griffiths and Hill, 1985) or using subroutines available in libraries such as NAG and IMSL. In any event, it is worth trying several different initial values to start an iterative scheme in order to check the stability of the numerical results.

Of course, even when regularity conditions on the likelihood function are satisfied, the asymptotic results may give poor approximations in small samples or samples with heavy censoring. In cases of doubt one can either search for exact distribution theory or adopt a more pragmatic approach and use simulation to examine the distribution of any appropriate estimator or test statistic; see Morgan (1984).

Other references of interest are DiCiccio (1987) and Lawless (1980) for the generalized gamma distribution and Shier and Lawrence (1984) for robust estimation of the Weibull distribution. Cox and Oakes (1984, Chapter 3) give a good general discussion of parametric methods for survival data. Johnson and Kotz (1970) give details of a whole range of estimation methods for the basic distributions discussed here. The paper by Lawless (1983), together with the resulting discussion is also well worth reading.

3.6 GOODNESS-OF-FIT

As part of a statistical analysis which involves fitting a parametric model, it is always advisable to check on the adequacy of the model. One may use either a formal goodness-of-fit test or appropriate data analytic methods. Graphical procedures are particularly valuable in this context.

Before discussing graphical methods for model checking, we mention formal goodness-of-fit tests. One approach is to embed the proposed model in a more comprehensive model. Tests such as those outlined in section 3.4 may then be applied. For example, one might test the adequacy of a exponential model relative to a Weibull model, as in Example 3.2 (continued). In contrast, the proposed model may be tested against a general alternative. This is the classical goodness-of-fit approach. An example is the well-known Pearson χ^2-test. The literature on goodness-of-fit tests is vast, though the tendency has been to concentrate on distributions such as the Normal, exponential and Weibull. A key reference is D'Agostino and Stephens (1986), but see also Lawless (1982, Chapter 9). Within the reliability context, however, tests based on the methods of section 3.4, combined with appropriate graphical methods will almost always be adequate.

In order to discuss graphical methods for checking model adequacy we shall use the notation introduced in section 2.11. That is, we suppose there are k distinct times $a_1 < a_2 < \cdots < a_k$ at which failures occur. Let d_j be the number of failures at time a_j and let n_j be the number of items at risk at a_j. In addition we shall use the plotting positions

$$p_j = 1 - \tfrac{1}{2}\{\hat{S}(a_j) + \hat{S}(a_j + 0)\},$$

where $\hat{S}$ is the PL estimator as in equation (2.20).

In section 2.11 we introduced a plotting procedure for models that depend only on location and scale parameters, μ and σ say, such as Gumbel and Normal distributions. In reliability this amounts to plotting the points

$$(\log a_j, F_0^{-1}(p_j)),$$

where F_0 is the distribution function of the proposed model with μ and σ set to 0 and 1 respectively. If the model is appropriate, the plot should be roughly linear. This type of plot, called the quantile–quantile (QQ) plot can be applied before a formal statistical analysis is attempted. In addition, rough parameter estimates for the proposed model may be obtained from the slope and intercept of the plot. These estimates may be of interest in their own right or may be used as starting values in an iterative scheme to obtain ML estimates. However, the applicability of this type of plot is limited. For example, it cannot be used for the gamma model. Moreover, the points on the plot which usually have the greatest visual impact, the extreme points, are those with the greatest variability.

A different graphical display, but which uses the same ingredients as the QQ plot, is the probability (PP) plot. This involves plotting the points

$$(p_j, F(a_j; \hat{\theta})),$$

where $F(a_j; \hat{\theta})$ denotes the distribution function of the proposed model, evaluated at the point a_j and with the parameters of the model set to 'reasonable' estimates (usually the ML estimates). Again, linearity of the plot is indicative of a good agreement between fitted model and data. Since estimates are used, the PP plot can only usually be constructed after fitting the model. However, its use is not limited to models with location and scale parameters only.

In the PP plot the extreme points have the lowest variability. A refinement which approximately stabilizes the variability of the plotted points is the stabilized probability (SP) plot; see Michael (1983). This involves plotting the points

$$\left(\frac{2}{\pi} \sin^{-1}\left(p_j^{1/2} \right), \frac{2}{\pi} \sin^{-1}\left\{ F^{1/2}\left(a_j; \hat{\theta} \right) \right\} \right).$$

Alternatively, simulated envelopes may be used to aid interpretation of QQ and PP plots; see Atkinson (1985). The construction of formal goodness-of-fit tests which are based on plots has also been investigated by Michael (1983), Kimber (1985) and Coles (1989).

Examples 3.3 (Example 2.3 continued)

The Weibull QQ plot for the strengths of 48 pieces of cord is shown in Figure 2.17. The overall impression of this plot is that it is basically linear except for the three smallest observed strengths. We now investigate the PP and SP plots for these data for a Weibull model.

Maximum likelihood estimation for a Gumbel distribution applied to the log-strengths gives $\hat{\mu} = 4.026$, and $\hat{\sigma} = 0.0613$ with variance-covariance matrix for the parameter estimates $(\hat{\mu}, \hat{\sigma})$

$$10^{-4}\begin{pmatrix} 0.9985 & -0.2191 \\ -0.2191 & 0.5889 \end{pmatrix}. \tag{3.10}$$

The PP and SP plots are shown in Figures 3.2 and 3.3 respectively. The visual impact of the three smallest points is much less than in Fig. 2.17. Overall the satisfactory fit of the Weibull model is confirmed.

Returning to the original purpose of analyzing these data, we wish to estimate $S(53)$. In Example 2.3 we found that the PL estimate is 0.685 with a

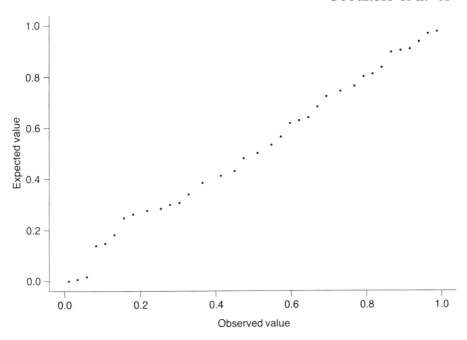

Figure 3.2 Weibull *PP* plot for the cord strength data.

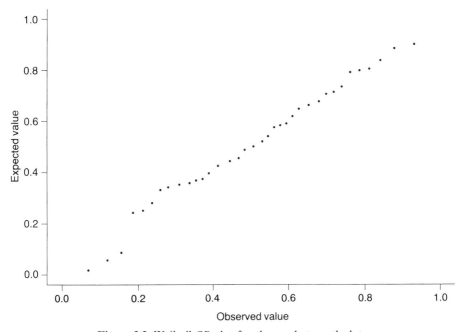

Figure 3.3 Weibull *SP* plot for the cord strength data.

standard error of 0.073. Using the Weibull model, the ML estimate of $S(53)$ is

$$\exp\left\{-\exp\left(\frac{\log 53 - 4.026}{0.0613}\right)\right\} = 0.668.$$

Using equation (3.3) in conjunction with the above variance-covariance matrix in expression (3.10), the standard error of the ML estimate is 0.046. This leads to a somewhat shorter approximate 95% confidence interval, [0.58, 0.76], than that obtained using the PL estimate, which was [0.54, 0.83].

4

Regression models for reliability data

4.1 INTRODUCTION

So far, we have considered reliability data in the context of a single experiment or observational study producing independent, identically distributed observations, some of which may be censored. In many applications of reliability, however, our real interest is in determining the way in which strength or lifetime depends on other variables, some of which may be under the operator's control. One example of this is the way in which failure time depends on the load applied to a component. It is almost always the case that increasing the load reduces the time to failure, but it is of interest to have a specific model for this. In this context, load is a covariate or explanatory variable. Other examples of possible covariates include temperature and the size of the specimens under study.

Covariates may be continuous or discrete. The three already mentioned are examples of continuous covariates because load, temperature and size are all variables that could in principle take any positive value. Discrete covariates arise when there are only a finite number of options. For example, in an experiment conducted with one of several different machines, if there were suspicion that the machines themselves are responsible for some of the differences between experiments, then it would make sense to treat the machine as a covariate.

The purpose of this chapter and the next is to present a range of techniques for dealing with data involving covariates. In view of the widespread use of the Weibull distribution is reliability and survival data analysis, there is particular interest in models based on this distribution. To some extent, these correspond to the classical models for regression and analysis of variance based on the Normal distribution. Indeed, an alternative 'lognormal' analysis would be to apply Normal-theory techniques to the logged observations. However, the Weibull analysis has the advantage of dealing more easily with censored data, quite apart from the theoretical features of the Weibull distribution, such as the fact that it is an extreme value distribution, and the fact that its hazard function is more reasonable for survival analysis than the hazard function of a lognormal distribution. These features make it often more appropriate than the lognormal distribution.

Initially, we shall assume that the variable of interest is a failure time T, though the same concepts apply equally well to failure load or other measures of reliability. Suppose the covariates are represented quantitatively by a vector x. For example, these might include the initial quality grade of the unit, operating stress, or maintenance support level. The purpose is to determine how T varies with x using an appropriate model for the dependence of T–probabilities on x and relevant data $(t_1, x_1), (t_2, x_2), \dots$. For example, it may be that the probabilities of longer lifetimes T increase with certain components of x. The model will be constructed to describe the nature of such changes and the data will include observations with x-vectors covering the range of interest. The data may then be used to estimate the unknown parameters of the model, to appraise the fit of the model and if necessary to suggest modifications of the model. Sections 4.2–7 are concerned with describing some commonly used models. The remaining sections then describe applications of these and related models to several real data examples.

4.2 ACCELERATED LIFE MODELS

Suppose we observe failure times for a system subject to a range of loads. One model that is easy to understand is that the failure time of a unit is the product of some function of the load, and a standardized failure time that represents the time at which the unit would have failed if the load had been at a specified standard level. This is known as an accelerated life model because of the interpretation that increasing the load has the effect of accelerating the rate of failure by a deterministic constant.

More formally, the accelerated life model holds when the survivor function $S(t; x) = \Pr\{T \geq t | x\}$ is of the form

$$S(t; x) = S_0(t\psi_x)$$

where S_0 is some baseline survivor function and ψ_x is a positive function of x. The interpretation is that, compared with some standardized system in which $\psi_x = 1$, the lifetime is divided by a factor ψ_x, or equivalently, the decay of the system is accelerated by a factor ψ_x. The limiting case $\psi_x = 0$ represents the situation where there is no stress and no failure, so $S(t; x) = 1$ for all t.

Example 4.1
Let T denote the time to breakdown of an electrical insulator subjected to constant voltage x, and suppose $\psi_x = x^\beta$ for some $\beta > 0$ and $S_0(t) = e^{-t}$. Then

$S(t, x) = \exp(-tx^\beta)$, so T is exponentially distributed with mean $x^{-\beta}$. Note that $\psi_0 = 0$ here.

In the accelerated life model the scaled lifetime $T\psi_x$ has survivor function $S_0(t)$ for all x. Equivalently, the log lifetime may be expressed as $\log T = -\log \psi_x + \log W$ where W has survivor function S_0. Thus the distribution of $\log T$ is just that of $\log W$ subject to a location shift.

The hazard function satisfies $h(t; x) = \psi_x h_0(t\psi_x)$ where h_0 is the baseline hazard function derived from S_0. This follows from

$$h(t; x) = -\frac{\partial}{\partial t} \log S(t; x) = -\frac{\partial}{\partial t} \log S_0(t\psi_x) = \psi_x h_0(t\psi_x).$$

As an initial guide to whether an accelerated life model is appropriate, a number of different plots may be tried.

1. Plots of observed log times against x can provide information on the form of ψ_x and can yield rough estimates of its parameters. Since $\log T = -\log \psi_x + \log W$, a plot of $\log T$ against x is effectively a plot of $-\log \psi_x$ against x, ignoring the 'error term' $\log W$.
2. Suppose that the data are grouped, the jth group of n_j observed failure times having $\psi_x = \psi_j$ and n_j being large enough for a plot of the sample survivor function $\hat{S}_j(t)$ to be meaningful. Then the plots of $\hat{S}_j(t)$ against $\log t$ should resemble horizontally shifted copies of the baseline plot of $S_0(t)$ against $\log t$. This follows from

$$S_j(t) = S_0(t\psi_j) = S_0\{\exp(\log \psi_j + \log t)\} = s_0(y + \log \psi_j)$$

say, with $y = \log t$. Here the function s_0 represents the survivor function of the log failure time when $\psi_j = 1$.
3. Another aspect of $S_j(t) = S_0(t\psi_j)$ is that the quantiles of T in the different groups are proportional. That is, if $q_j(p) = S_j^{-1}(p)$ is the pth quantile of T in group j, then $q_j(p)\psi_j = q_k(p)\psi_k$ (since $S_0\{q_j(p)\psi_j\} = S_j\{q_j(p)\} = p$ for each j). These quantiles may be estimated from the $\hat{S}_j(t)$–plots in (2.), say for $p = 0.1, 0.2, ..., 0.9$; thus one obtains estimates $\hat{q}_{j1}, \hat{q}_{j2}, \cdots \hat{q}_{j9}$ for each group j. A plot of the points $(\hat{q}_{11}, \hat{q}_{j1}), ..., (\hat{q}_{19}, \hat{q}_{j9})$, taking group 1 quantiles along the x axis, should resemble a straight line of slope ψ_1/ψ_j through the origin. Alternatively, a similar plot of log quantiles should yield a set of parallel straight lines of slope 1. It may be worth replacing the abscissae of the plot, here suggested as those from group 1, by the average values over the groups; this will avoid erratic values, but require more computation.
4. The above plots apply to censored as well as uncensored data. In the case of censored data, the Kaplan–Meier estimator is used for $\hat{S}$.

4.3 PROPORTIONAL HAZARDS MODELS

In the proportional hazards models the hazard function satisfies $h(t; x) = h_0(t)\psi_x$. Here $h(t; x)$ is the hazard under factors x, $h_0(t)$ is some baseline hazard function, and ψ_x is some positive function of x. Since $h(t; x_1)/h(t; x_2) = \psi_{x_1}/\psi_{x_2}$ is independent of t, the hazards at different x values are in constant proportion over time, hence the name. Provided $\psi_0 > 0$, there is no loss of generality in assuming $\psi_0 = 1$ since one can redefine $h(t; x) = h_0(t)\psi_x/\psi_0$.

The model continues to hold under monotone transformation of the time scale. Suppose that s is an increasing or decreasing function of t, e.g. a transformed scale such as $s = \log t$. Then, defining $h^*(s; x)$ and $h_0^{\#}$ to be hazard functions on the s scale, we have $h(t; x) = h^*(s; x) \, ds/dt$, $h_0(t) = h_0^{\#}(s) \, ds/dt$ and hence $h^*(s; x) = h_0^{\#}(s)\psi_x$, a proportional hazards model with the same ψ function.

The survivor function satisfies $S(t; x) = S_0(t)^{\psi_x}$ for this model, S_0 being the baseline survivor function corresponding to h_0. This is because

$$\log S(t; x) = -\int_0^t h(t; x) \, dt = -\psi_x \int_0^t h_0(t) \, dt = \psi_x \log S_0(t).$$

The family $\{S_0(t)^{\psi} : \psi > 0\}$ of survivor functions is called the class of Lehmann alternatives generated by $S_0(t)$.

Suppose that $S(t; x)$ is both an accelerated life and a proportional hazards model, so that

$$S(t; x) = S_{a0}(t\psi_{ax}) = S_{p0}(t)^{\psi_{px}},$$

with the baseline functions standardized so that $\psi_{a0} = \psi_{p0} = 1$. Then

$$S_{a0}(t) = S_{p0}(t) = \exp(-t^{\eta})$$

for some $\eta > 0$ (Weibull survivor function) and $\psi_{px} = \psi_{ax}^{\eta}$. For a proof see Cox and Oakes (1984, section 5.3).

Again, it is worthwhile to begin the analysis by plotting the data. Suppose that the data are grouped as in (2.) of section 4.2. Then the plots of $-\log \hat{S}_j(t)$ against t for the different groups should resemble multiples of the baseline curve $-\log S_0(t)$ against t; for, if $S_j(t) = S_0(t)^{\psi_j}$ then $-\log S_j(t) = -\psi_j \log S_0(t)$. Alternatively, the plots of $\log\{-\log \hat{S}_j(t)\}$ against t should resemble vertically shifted copies of $\log\{-\log S_0(t)\}$ against t. The t axis may be transformed, e.g. by taking logarithms, without changing the pattern of vertically parallel curves. This may yield a tidier plot, as illustrated in section 4.8 below.

The proportional hazards model is developed in more detail in Chapter 5.

4.4 PROPORTIONAL ODDS MODELS

In the proportional odds model the survivor function satisfies

$$\{1 - S(t; x)\}/S(t; x) = \psi_x\{1 - S_0(t)\}/S_0(t) \tag{4.1}$$

(Bennett, 1983). The defining ratio $\{1 - S(t; x)\}/S(t; x)$ is $\Pr\{T < t\}/\Pr\{T \geq t\}$ which is the 'odds' on the event $T < t$. Thus equation (4.1) says that the odds under explanatory variables x is ψ_x times the baseline odds, hence the name. As ψ_x increases, so does the probability of a shorter lifetime. As with proportional hazards, the model continues to hold under transformation of the time scale.

Differentiating equation (4.1) with respect to t yields $h(t; x)/S(t; x) = \psi_x h_0(t)/S_0(t)$. Hence, the hazard ratio satisfies

$$h(t; x)/h_0(t) = \psi_x S(t; x)/S_0(t) = \{1 - S(t; x)\}/\{1 - S_0(t)\},$$

the second equality following from reapplication of equation (4.1). It follows that the hazard ratio is ψ_x at $t = 0$, and tends to 1 as $t \to \infty$. This behaviour reflects a diminishing effect of x on the hazard as time goes on: either the system adjusts to the factors imposed on it, or the factors operate only in the earlier stages.

To construct diagnostic plots for this model, suppose that the data are grouped as in (2.) of section 4.2. Then the plots of $\{1 - \hat{S}_j(t)\}/\hat{S}_j(t)$ against t for the different groups should resemble multiples of the corresponding baseline odds curve, and hence multiples of each other. Equivalently, plots of $\log[\{1 - \hat{S}_j(t)\}/\hat{S}_j(t)]$ should be vertically shifted copies of the baseline curve, and the horizontal scale can be transformed, e.g. to $\log t$, for a tidier plot.

4.5 GENERALIZATIONS

Lawless (1986) described generalized versions of the models treated above. For the proportional hazards model, he replaced the relation $\log\{-\log S_j(t)\} = \log\{-\log S_0(t)\} + \log \psi_j$ (used above for the plotting) by $\phi_1\{S_j(t)\} = \phi_1\{S_0(t)\} + g_1(\psi_j)$ where ϕ_1 and g_1 are specified functions. For the accelerated life model the quantile relation $\log q_j = \log q_0 - \log(\psi_j/\psi_0)$ is replaced by $\phi_2(q_j) = \phi_2(q_0) - g_2(\psi_j)$. The proportional odds model is covered by taking $\phi_1(s) = \log\{(1 - s)/s\}$ in the first generalization. Lawless discussed procedures for plotting and data analysis associated with these models, and also identified models which satisfy both generalized equations.

Some particular generalizations which have been proposed in the literature are as follows.

1. Generalized accelerated life: $S(t; x) = S_0\{\psi_x(t)\}$ where $\psi_x(t)$ increases

monotonically from 0 to ∞ with t, e.g. $\psi_x(t) = t\psi_x$, t^{ψ_x} or $\exp(t\psi_x)$ with $\psi_x > 0$.

2. Generalized proportional hazards: $h^{(\lambda)}(t; x) = h_0^{(\lambda)}(t) + \psi_x^{(\lambda)}$ where $h^{(\lambda)} = (h^\lambda - 1)/\lambda$ for $\lambda \neq 0$, $h^{(0)} = \log h$, the Box–Cox (1964) transformation of h.

3. Time shift: $h(t; x) = h_0(t + \psi_x)$; this represents a delayed-action hazard.

4. Polynomial hazard: $h(t; x) = \psi_{0x} + \psi_{1x}t + \psi_{2x}t^2 + \cdots + \psi_{qx}t^q$.

5. Multiphase hazard: $h(t; x) = h_j(x; t)$ for $t_{j-1} < t < t_j$ (jth phase, $j = 1, \ldots, m$) with $t_0 = 0, t_m = \infty$. Here there are m phases, e.g. the first with high hazard representing early failure of defective units.

6. Time-dependent covariates: $x = x(t)$; here the operating stress varies over time.

All these models may, of course, be applied to other variables, such as failure strength, as well as failure time.

4.6* AN ARGUMENT FROM FRACTURE MECHANICS

In this section we outline an argument based on fracture mechanics which suggests, in the absence of evidence to the contrary, that for problems associated with failure time under fatigue, there may be reasons for considering the accelerated life model in preference to the others in this chapter.

A standard model for fatigue crack growth may be expressed in the form

$$\frac{da}{dt} = K_1(a^{1/2}L)^\beta \tag{4.2}$$

where $a = a(t)$ is the length of a crack at time t, $L = L(t)$ is the stress amplitude at time t and $K_1 > 0, \beta \gg 2$ are constants depending on the material and the 'geometry' of the specimen. Solving equation (4.2) yields

$$a_0^{1-\beta/2} - a^{1-\beta/2} = K_1(\beta/2 - 1) \int_0^t \{L(s)\}^\beta \, ds$$

where $a_0 = a(0)$ is the initial crack length. Failure occurs when $a(t)$ reaches a critical crack length a_c, which depends on the applied stress L through a relation of the form $a_c = K_2 L^{-2}$. This is derived from the well-known Griffith theory of brittle fracture, which predicts that the critical stress L for a crack of length a is proportional to $a^{-1/2}$. Inverting this yields that critical crack length depends on L proportionately to L^{-2}. Failure then occurs by time t if, for some $\tau \leq t$, we have

$$a(\tau)^{1-\beta/2} = a_0^{1-\beta/2} - K_1(\beta/2 - 1) \int_0^\tau L^\beta(s) \, ds < K_2^{(2-\beta)/2} L(\tau)^{\beta-2},$$

in other words if

$$a_0^{1-\beta/2} < \sup_{\tau \le t} \left[K_1(\beta/2-1) \int_0^\tau L^\beta(s)\,ds + K_2^{(2-\beta)/2} L(\tau)^{\beta-2} \right]. \qquad (4.3)$$

The failure time of the specimen is obtained by substituting A_{max} for a_0 in (4.3), where A_{max} is the (random) largest crack length in the system. The reasoning behind this is simply that, for a homogeneous material, the longest initial crack is the one that will fail first under fatigue loading.

Since A_{max} is an extreme value, it is natural to adopt one of the extreme value distributions for it. The most convenient is the Fréchet form

$$\Pr\{A_{max} \le a\} = \exp\{-(a/a^*)^{-\gamma}\} \quad (a^* > 0, \gamma > 0).$$

Identifying a_0 with A_{max}, it then follows that

$$\begin{aligned}
\Pr\{a_0^{1-\beta/2} > y\} &= \Pr\{a_0 < y^{-2/(\beta-2)}\} \\
&= \exp[-(a^*)^\gamma y^{2\gamma/(\beta-2)}] \\
&= \exp[-(y/y^*)^\eta] \quad \text{say}, \quad \text{defining } \eta = 2\gamma/(\beta-2).
\end{aligned}$$

We then have

$$\Pr\{T \le t\} = G\left[\sup_{\tau \le t} \left\{ K_1(\beta/2-1) \int_0^\tau L^\beta(s)\,ds + K_2^{(2-\beta)/2} L(\tau)^{\beta-2} \right\} \right] \qquad (4.4)$$

where $G(y) = 1 - \exp[-(y/y^*)^\eta]$.

This is the most general form obtainable from this argument; we have given it in full because it is referred to again in section 9.8. In most fatigue-related applications we are interested in large failure times and stresses L which are small relative to the stress-rupture failure stress on the specimen. In that case the K_2 term in equation (4.4) may be neglected and we obtain the much simpler expression

$$\Pr\{T \le t\} = G\left[K_1(\beta/2-1) \int_0^t L^\beta(s)\,ds \right]$$

which when L is constant simplifies further to

$$\Pr\{T \le t\} = 1 - \exp[-\{K_1(\beta/2-1)tL^\beta/y^*\}^\eta] = 1 - \exp\{t/t^*)^\eta\} \qquad (4.5)$$

where $t^* = 2y^* K_1^{-1} L^{-\beta}/(\beta-2)$. This form has been widely used, e.g. Phoenix (1978).

An alternative approach would be to assume that the variation in failure times derives from variation in the constant K_1 from specimen to specimen. An approach which has been used for aircraft lifing purposes is to determine a 'maximum' initial crack size, which might represent the limits of the manufacturing process and/or non-destructive inspection capability. One then assumes that a crack of this size is initially present at a critical location in the specimen, and K_1 is regarded as a random quantity. Since T is proportional to K_1, a lognormal distribution for K_1 will imply a lognormal distribution for failure time. Alternatively, a Fréchet form for K_1 leads to a Weibull distribution for T, as above, though in this case one cannot appeal to extreme value theory to support such a conclusion.

In summary, considerations from fracture mechanics suggest an accelerated life law, with equation (4.5) suggested as a specific model. This model is based on a number of assumptions, however, the most critical for the accelerated life argument being that the randomness of failure time is primarily a consequence of random cracks in the material at the start, crack growth being an essentially deterministic process. This is a widely held view but is by no means universal; for example Kozin and Bogdanoff (1981) have reviewed models of stochastic crack growth, and this is still a question of active research.

4.7 MODELS BASED ON THE WEIBULL DISTRIBUTION

As mentioned in the introduction to this chapter, there is particular interest in regression models based on the Weibull distribution. For the remainder of the chapter, we shall concentrate primarily on this distribution, showing in particular how many of the concepts for regression and analysis of variance using the Normal distribution and least squares, may also be applied using the Weilbull distribution and numerical maximum likelihood fitting.

One simple form of Weibull model uses the survivor function

$$S(t; x) = \exp\{-(t/\alpha_x)^\eta\} \tag{4.6}$$

where we may take $\log \alpha_x = x^T \beta$ (loglinear model) and η independent of x. This is a fully parametric model with parameters β (regression coefficient) and η (Weibull shape parameter). The hazard function is $h(t; x) = \eta t^{\eta-1}/\alpha_x^\eta$. Let $S_0(t) = \exp(-t^\eta)$. Then $S(t; x)$ can be expressed as $S_0(t\psi_{ax})$ with $\psi_{ax} = 1/\alpha_x$, an accelerated life model, or $S_0(t)^{\psi_{px}}$ with $\psi_{px} = \alpha_x^{-\eta}$, a proportional hazards model. Alternatively, we may represent the model in log form by

$$\log T = \log \alpha_x + W/\eta \tag{4.7}$$

where W has a Gumbel distribution with survivor function $\exp(-e^w)$ on

$[-\infty, \infty]$. This follows from

$$\Pr\{W \geq w\} = \Pr\{\log(T/\alpha_x) \geq w/\eta\} = S(\alpha_x e^{w/\eta}; x) = \exp(-e^w).$$

Thus, from section 3.4, $\log T$ has mean $\log \alpha_x - \gamma/\eta$ (where γ is Euler's constant) and variance $\pi^2/(6\eta^2)$. This representation also makes clear the role of $1/\eta$ as a scale constant for $\log T$.

Within the framework of parametric models, there are a number of more general possibilities which could be considered. One possibility is to extend equations (4.6) and (4.7) in the form

$$S(t; x) = \exp\{-(t/\alpha_x)^{\eta_x}\}, \quad \log T = \log \alpha_x + W/\eta_x \qquad (4.8)$$

so that η as well as α depends on x. This may be appropriate when initial exploratory analysis of the data suggests that the Weibull shape parameter is not the same everywhere.

There are also possible extensions to involve the three-parameter Weibull or generalized extreme value distributions (recall section 2.7). For example, equation (4.8) may be extended to

$$S(t; x) = \exp[-\{(t - \theta_x)/\alpha_x\}^{\eta_x}] \qquad (4.9)$$

where any of θ_x, α_x or η_x may be constant or may depend on x. Similarly, but not equivalently, the log-representation in equation (4.8) may be extended to give

$$S(t; x) = \exp[-\{1 - \gamma_x(\log t - \mu_x)/\sigma_x\}^{-1/\gamma_x}], \qquad 0 < t < \infty, \qquad (4.10)$$

in which the limiting case $\gamma_x \to 0$ is equivalent to equation (4.7) but in general the distribution of $\log T$ is extended from Gumbel to generalized extreme value. The motivation for using equation (4.10) as opposed to equation (4.9) is partly based on the knowledge that the generalized extreme value parametrization is more stable than the traditional three-parameter Weibull parametrization (Smith and Naylor 1987).

Estimation and testing
For simple linear regression, $\log \alpha_x = \beta_0 + \beta_1 x$, with uncensored data, a plot of the observed $\log T$ against x should be a straight line of slope β_1 and intercept $\beta_0 - \gamma/\eta$. The variance about this line is $\pi^2/(6\eta^2)$. Thus an ordinary least squares fit will yield estimates of β_0, β_1 and η, though these are theoretically inefficient compared with maximum likelihood estimates. Lawless (1982) made detailed comparisons between least squares and maximum likelihood fits for models of this form. In the multiple regression case $\log \alpha_x = x^T \beta$, one can

similarly apply least squares methods to estimate β and η, and to make rough checks of model assumptions as in Chapter 3.

More generally, the method of maximum likelihood may be applied. The principles of this are identical to those described in Chapter 3: a log likelihood function may be written in the form

$$l(\phi) = \sum_{i \in U} \log f(t_i; x_i, \phi) + \sum_{i \in C} \log S(t_i; x_i, \phi)$$

where U is the set of uncensored and C the set of (right-) censored observations. Here, ϕ is the vector of all unknown parameters (consisting, for example, of β and η in the simple loglinear case just described), f is the density function corresponding to survivor function S, and t_i, x_i, are respectively the T and x values for the ith observation. More complicated forms of censoring, such as interval censoring, can be handled similarly, as described in Chapter 3. The maximum likelihood estimator is then that value $\hat{\phi}$ which maximizes $l(\phi)$, and the set of second-order derivatives of $-l$, evaluated at $\hat{\phi}$, is the **observed information matrix** for ϕ; its inverse is an approximation to the variance-covariance matrix for $\hat{\phi}$. Standard errors for $\hat{\phi}$ are the square roots of the diagonal elements of the estimated variance-covariance matrix. To test between nested models M_0 and M_1 say, where M_0 with p_0 parameters is contained within M_1 which has $p_1 > p_0$ parameters, the likelihood ratio statistic $W = 2(l_1 - l_0)$ has an approximate $\chi^2_{p_1 - p_0}$ distribution. Here l_1 and l_0 are the maximized values of l under M_1 and M_0 respectively. These results are all justified by standard large-sample theory under regularity conditions (recall section 3.4). In the present context the main regularity condition of importance is that $\gamma > -0.5$ in the generalized extreme value distribution, or equivalently $\eta > 2$ in the three-parameter Weibull distribution. Alternative procedures when $\gamma \leq -0.5$ or $\eta \leq 2$ have been proposed (Smith 1985, 1989) but this is a more advanced topic which will not be pursued here. For the simpler forms of the models given in equations (4.6) and (4.7), maximum likelihood methods are always regular in the sense that the standard large-sample procedures are valid.

As in the i.i.d. case, computation may be based on Newton–Raphson or quasi-Newton iterative methods. Alternatively, it is possible to fit these models by the use of suitable macros in GLIM. In the case of the generalized extreme value distribution without covariates, a Newton–Raphson algorithm is described by Prescott and Walden (1983) and a Fortran algorithm is given by Hosking (1985) (with an amendment by Macleod, 1989). For problems with covariates these algorithms are easily generalized, or alternatively it is possible to use packaged optimization routines of the sort readily available in numerical libraries such as NAG or IMSL. The main proviso is the importance of using reasonable starting values. These may be obtained either by doing an initial least-squares fit, or else by adding one new parameter at a time to the model, using the results of previous fits to guess starting values.

Estimation of other quantities of interest, such as population quantiles, may proceed by substitution of the maximum likelihood estimates, with standard errors calculated by the delta method exactly as described in section 3.2.

Residual plots
Of the various methods of assessing the fit of an assumed model, the main one developed here is a graphical method based on probability plots of generalized residuals, generalizing the PP method of section 3.6. Once a model has been fitted, the transformation

$$u_i = S(t_i; x_i, \hat{\phi}) \qquad (4.11)$$

reduces the distribution of the ith observation, assuming it is uncensored, to uniformity on [0, 1], so plots and tests of fit may be based on the u_i. This idea of forming generalized residuals is due to Cox and Snell (1968); they also proposed a bias-correction to adjust for the estimation of ϕ, but in most cases they are used in their unadjusted form. For our present purposes, given that so much of the discussion is in terms of the Gumbel distribution, we transform again to

$$z_i = \pm \log(-\log u_i), \qquad (4.12)$$

(making it a QQ plot) where taking the $+$ sign corresponds to the survivor function $S(t) = \exp(-e^x)$ while the $-$ sign leads to the corresponding form of the Gumbel distribution appropriate for maxima rather than minima, $S(t) = 1 - \exp(-e^{-t})$. It is convenient to permit both forms so as to allow interchangeability of software between extreme-value maximum and extreme-value minimum applications.

Taking for the moment the $+$ sign in equation (4.12), a probability plot may be contructed by first ordering the z_i as $z_{1:n} \leq \cdots \leq z_{n:n}$ and then plotting $z_{i:n}$ against $\log\{-\log(1 - p_{i:n})\}$ where $p_{i:n}$ is an apropriate plotting position. Here we take $p_{i:n} = (i - 0.5)/n$ in the case where all observations are uncensored. In the right-censored case this may be replaced by a plotting position deduced from the Kaplan–Meier estimator

$$p_{i:n} = 1 - \frac{n + 0.5}{n} \prod_{j \leq i, j \in J} \frac{(n - j + 0.5)}{(n - j + 1.5)}$$

where J denotes the set of indices corresponding to uncensored observations; note that this is after reordering the z_i so we do not equate J with U given earlier. The idea of forming QQ plots in this way was developed extensively by Chambers *et al.* (1983).

If the model is a good fit, then the probability plot should be very close to the straight line through the origin of unit slope, and visual comparison of the two is often the best way of assessing the fit of a model. In the case when the original data are believed to follow the Gumbel distribution without covariates, the same plot may be applied to the original data, and it then provides a good initial indication of the fit of the Gumbel distribution as well as suggesting initial estimates of the parameters. In this case the term **reduced values** is also used to describe the $\log\{-\log(1-p_{i:n})\}$. The two-parameter Weibull distribution is handled the same way, after taking logarithms.

Plots of generalized residuals may be used in a number of ways. For example, as well as doing a probability plot as just described, they may also be plotted against individual covariates. Another idea is to extract a subset of residuals according to the value of some covariate or group of covariates, and then to form a separate probability plot for that subset. The effectiveness of this idea will be seen in section 4.10.

4.8 AN EXAMPLE: BREAKING STRENGTHS OF CARBON FIBRES AND BUNDLES

Table 4.1 gives failure stresses (in GPa) of four samples of carbon fibres, of lengths, 1, 10, 20 and 50 mm, while Table 4.2 gives corresponding failure stresses for bundles of 1000 fibres, of varying lengths, embedded in resin. The data are taken from experiments conducted in the Department of Materials Science and Engineering, University of Surrey, and have previously been discussed by Watson and Smith (1985) and Lindgren and Rootzén (1987).

First, we illustrate the various plots discussed in section 4.2–4 for the single-fibre data, the explanatory variable here being the length of the fibre. Under the accelerated life model, the distributions of $\log T$ for the different lengths are location-shifted versions of one another. One consequence is that the central moments of $\log T$, such as the variance, in different data sets should be equal. The sample variances of the $\log T$'s at the four lengths are 0.042, 0.040, 0.045 and 0.037. These values seem close and Bartlett's test for homogeneity of variances (see e.g. Kennedy and Neville 1986, Chapter 16) yields a χ_3^2 value of 0.73, clearly insignificant.

Figure 4.1 shows plot (2.) of section 4.2. The curves should look like horizontally shifted copies of one another under the model. The pattern is marginally acceptable to an optimistic observer.

For plot (3.) of section 4.2, the log-quantiles are first estimated. The p-quantile of a sample of size n was taken to be the ith order statistic, where $i = np + 0.5$; note that this corresponds to the plotting position formula $p_{i:n} = (i - 0.5)/n$. In the case where this i is not an integer, the value taken is a

Table 4.1 Failure stresses of single carbon fibres

(a) Length 1 mm

2.247 2.64 2.842 2.908 3.099 3.126 3.245 3.328 3.355 3.383 3.572 3.581 3.681
3.726 3.727 3.728 3.783 3.785 3.786 3.896 3.912 3.964 4.05 4.063 4.082 4.111
4.118 4.141 4.216 4.251 4.262 4.326 4.402 4.457 4.466 4.519 4.542 4.555 4.614
4.632 4.634 4.636 4.678 4.698 4.738 4.832 4.924 5.043 5.099 5.134 5.359 5.473
5.571 5.684 5.721 5.998 6.06

(b) Length 10 mm

1.901 2.132 2.203 2.228 2.257 2.35 2.361 2.396 2.397 2.445 2.454 2.454 2.474
2.518 2.522 2.525 2.532 2.575 2.614 2.616 2.618 2.624 2.659 2.675 2.738 2.74
2.856 2.917 2.928 2.937 2.937 2.977 2.996 3.03 3.125 3.139 3.145 3.22 3.223
3.235 3.243 3.264 3.272 3.294 3.332 3.346 3.377 3.408 3.435 3.493 3.501 3.537
3.554 3.562 3.628 3.852 3.871 3.886 3.971 4.024 4.027 4.225 4.395 5.02

(c) Length 20 mm

1.312 1.314 1.479 1.552 1.7   1.803 1.861 1.865 1.944 1.958 1.966 1.997 2.006
2.021 2.027 2.055 2.063 2.098 2.14 2.179 2.224 2.24 2.253 2.27 2.272 2.274
2.301 2.301 2.339 2.359 2.382 2.382 2.426 2.434 2.435 2.478 2.49 2.511 2.514
2.535 2.554 2.566 2.57 2.586 2.629 2.633 2.642 2.648 2.684 2.697 2.726 2.77
2.773 2.8   2.809 2.818 2.821 2.848 2.88 2.954 3.012 3.067 3.084 3.09 3.096
3.128 3.233 3.433 3.585 3.585

(d) Length 50 mm

1.339 1.434 1.549 1.574 1.589 1.613 1.746 1.753 1.764 1.807 1.812 1.84 1.852
1.852 1.862 1.864 1.931 1.952 1.974 2.019 2.051 2.055 2.058 2.088 2.125 2.162
2.171 2.172 2.18 2.194 2.211 2.27 2.272 2.28 2.299 2.308 2.335 2.349 2.356
2.386 2.39 2.41 2.43 2.431 2.458 2.471 2.497 2.514 2.558 2.577 2.593 2.601
2.604 2.62 2.633 2.67 2.682 2.699 2.705 2.735 2.785 2.785 3.02 3.042 3.116
3.174

linear interpolation between the two nearest order statistics. The resulting plot of quantiles for groups 2, 3, 4 against those for group 1 are shown in Figure 4.2, and the plot of all four groups against the average quantiles is shown in Figure 4.3. The resemblence to a set of parallel lines, as required by the model, is not definitive, though better in Figure 4.3 than in Figure 4.2.

The proportional hazards model may be examined via a plot of the log-log survivor function against log t, as described in section 4.3. Figure 4.4 shows a plot of $\log(t_{i:n})$ against $\log(-\log(1-p_{i:n}))$ for each sample, where $t_{i:n}$ is the ith order statistic from a sample of breaking strengths of size n, and $p_{i:n} = (i-0.5)/n$. In this case, a plot consisting of vertically shifted copies of the same curve indicates a proportional hazards model, and one consisting of parallel straight lines implies a Weibull model as in equation (4.6). For comparision the four Weibull lines, as obtained by maximum likelihood fitting of a two-parameter Weibull distribution to each subsample, are superimposed on the plot. In this

Table 4.2 Failure stresses of bundles of 1000 impregnated carbon fibres

(a) Length 20 mm
2.526 2.546 2.628 2.628 2.669 2.669 2.71 2.731 2.731 2.731 2.752 2.752 2.793
2.834 2.834 2.854 2.875 2.875 2.895 2.916 2.916 2.957 2.977 2.998 3.06 3.06
3.06 3.08

(b) Length 50 mm
2.485 2.526 2.546 2.546 2.567 2.628 2.649 2.669 2.71 2.731 2.752 2.772 2.793
2.793 2.813 2.813 2.854 2.854 2.854 2.895 2.916 2.936 2.936 2.957 2.957 3.018
3.039 3.039 3.039 3.08

(c) Length 150 mm
2.11 2.26 2.34 2.44 2.51 2.51 2.57 2.57 2.61 2.61 2.61 2.65 2.67
2.71 2.71 2.71 2.75 2.75 2.75 2.75 2.77 2.77 2.79 2.83 2.83 2.83
2.87 2.87 2.9   2.9   2.92 2.94

(d) Length 300 mm
1.889 2.115 2.177 2.259 2.279 2.32 2.341 2.341 2.382 2.382 2.402 2.443 2.464
2.485 2.505 2.505 2.526 2.587 2.608 2.649 2.669 2.69 2.69 2.71 2.751 2.751
2.854 2.854 2.875

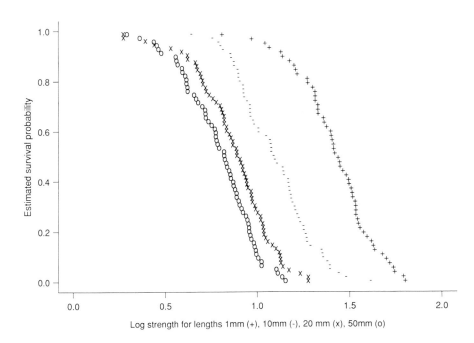

Figure 4.1 Plot of survival probability *vs.* log strength for single fibre data.

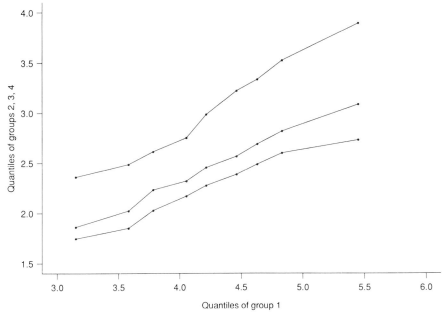

Figure 4.2 Quantile plot for single fibres data: quantiles of samples 2, 3, 4 (lengths 10, 20, 50 mm.) against quantiles of sample 1 (1 mm.).

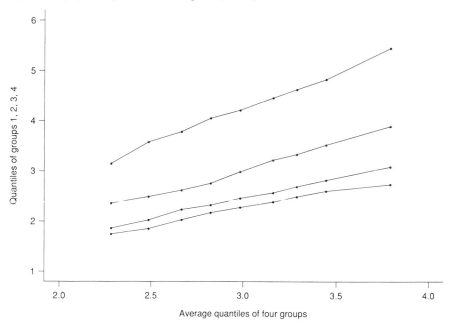

Figure 4.3 Quantile plot for single fibres data: quantiles of all four samples against average quantiles of the four samples.

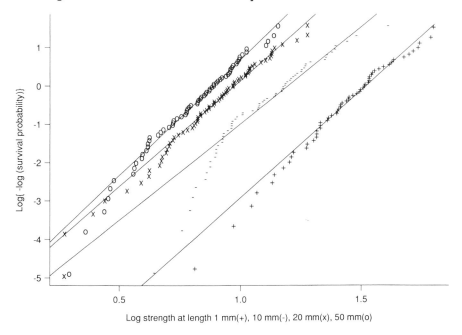

Figure 4.4 Weibull plots for four samples of single fibres data, with fitted Weibull lines estimated separately for each sample.

case the plots for 1, 20 and 50 mm are acceptably close to parallel straight lines, but that for 10 mm contains a definite curvature.

The proportional odds model may be tested by plotting the log odds against $\log t$ as described in section 4.4. The result is given in Figure 4.5. The required model, of vertically shifted copies, is not established, and indeed the curves for lengths 1 mm and 10 mm appear to be respectively concave and convex.

For the remainder of this section, we concentrate on the Weibull models, focusing particularly on the **weakest link hypothesis**. The notion that a long fibre consists of independent links, whose weakest member determines the failure of the bundle, leads to the model

$$S_L(t) = S_1(t)^L \tag{4.13}$$

relating the survivor function S_L for a fibre of length L to the survivor function S_1 for length 1. In the Weibull case this leads to model M_0:

$$S_L(t) = \exp\{-L(t/t_1)^\eta\} = \exp\{-(t/t_L)^\eta\} \tag{4.14}$$

where $t_L = t_1 L^{-1/\eta}$. As alternatives to this, we may consider:

Model M_1: as (4.14) but with $t_L = t_1 L^{-\xi/\eta}$ with $0 \leq \xi \leq 1$;

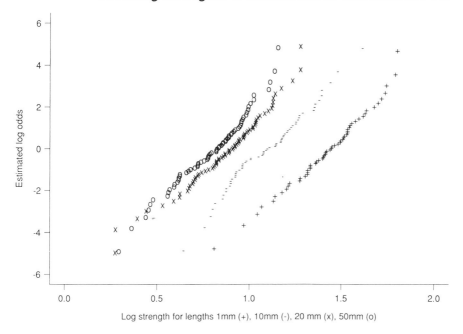

Log strength for lengths 1mm (+), 10mm (-), 20 mm (x), 50mm (o)

Figure 4.5 Proportional odds plot for single fibres data.

Model M_2: as M_1 but the $\eta = \eta_L$ dependent on L through a relation of the form $\log \eta_L = \log \eta_1 + \tau \log L$ (this is a log-linear model for the scale parameter of the Gumbel distribution for $\log T$);

Model M_3: separate two-parameter Weibull distributions for each length.

Each of these models is a special case of (4.8) and the parameters may be estimated by numerical maximum likelihood. The easiest of the four models to fit is of course model M_3, since in this case we merely have to fit a separate Weibull distribution to the four sub-samples using the methods of Chapter 3, the final log likelihood being the sum of the log likelihoods for the four sub-samples. To illustrate the technique for the others, consider M_1. If we have data $\{T_i, \delta_i, L_i, i = 1, \ldots, n\}$ where T_i is either the failure time or the (right-) censoring time of the ith specimen, δ_i is 1 if T_i is a failure time and 0 if it is a censoring time, and L_i is the length of the ith specimen, then the log likelihood is

$$l = \sum_{i=1}^{n} \delta_i \{\log \eta + (\eta - 1) \log T_i + \xi \log L_i - \eta \log t_1\} - t_1^{-\eta} \sum_{i=1}^{n} L_i^{\xi} T_i^{\eta}. \quad (4.15)$$

For fixed η and ξ, this can be maximized analytically for t_1, leading to the estimate

$$t_1 = \left\{ \frac{1}{r} \sum_{i=1}^{n} L_i^{\xi} T_i^{\eta} \right\}^{1/\eta} \quad (4.16)$$

where $r = \sum \delta_i$ is the number of uncensored observations. This estimate can then be substituted back into equation (4.15) to produce a log likelihood in terms of just η and ξ. In the case of model M_0, we just set $\xi = 1$ and perform a one-dimensional search (which can, if desired, be done graphically or by trial and error) to find $\hat{\eta}$. Even in the case of model M_1, it is possible to do the same thing for a sequence of trial values of ξ (say, 0 to 1 in steps of 0.1) and then to pick out an appropriate ξ amongst these. Model M_2 is rather harder, however, since in this case η in equation (4.15) must be replaced by $\eta_1 L_i^{\tau}$ which, being different for each L_i, does not allow the simple reduction to (4.16). This discussion is included, nevertheless, to illustrate the point that for models M_0, M_1 and M_3, a full-scale numerical optimization is not strictly necessary, since the same results can be obtained by more elementary techniques.

Once the models are fitted, they may be compared using the likelihood ratio criterion applied to maximized log likelihoods l_0, l_1, l_2, and l_3.

For the single fibres data we obtain

$l_0 = -229.1$ (2 parameters),
$l_1 = -227.7$ (3 parameters),
$l_2 = -227.6$ (4 parameters),
$l_3 = -220.1$ (8 parameters).

In this case the sharp increase to l_3 casts doubt on whether any of models M_0–M_3 is adequate, though Watson and Smith (1985) argued that the practical difference between the models is not great. The main interest, therefore, seems to lie in the distinction between M_0 and M_1, which is not significant as judged by the χ^2 test. The parameter ξ is of particular interest here, since this is equivalent to a size effect in which 'size' is proportional to L^{ξ} rather than L itself. For this data set $\hat{\xi} = 0.90$.

For the bundles data in Table 4.2, the model-fitting analysis has been repeated obtaining

$l_0 = 21.8$ (2 parameters),
$l_1 = 29.6$ (3 parameters),
$l_2 = 31.5$ (4 parameters),
$l_3 = 35.8$ (8 parameters).

The χ^2 tests seem to point to M_1 as the model of choice, with $\hat{\xi} = 0.58$, significantly different from 1. Figure 4.6 shows the proportional hazards plots for this data set.

One criticism of these methods as tests of the weakest link effect is that they depend on the Weibull distribution which, although widely used partly because it is consistent with the weakest link effect, is not actually part of the weakest link assumption. It would be better to test weakest link directly, without

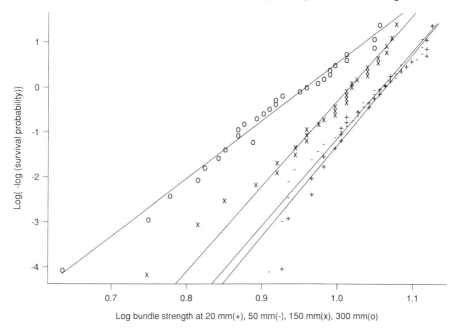

Figure 4.6 Weibull plots for four samples of impregnated bundles, with fitted Weibull lines.

assuming a Weibull distribution. To this end Lindgren and Rootzén (1987) proposed a test based just upon (4.13), without the Weibull assumption. The idea behind this is the very simple observation that (4.13) is equivalent to a proportional hazards hypothesis relating the failure stresses at different lengths. An entirely non-parametric procedure has been developed by Wolstenholme (1989), based purely on the order of failure in four sections of different lengths cut from the same fibre. The advantage of Wolstenholme's procedure is that it eliminates fibre-to-fibre variation. Neither of these techniques has been applied very widely so far, but results which have been reported contradict the weakest link hypothesis even more clearly than the existing parametric analyses.

   Given the wide acceptance of the weakest link notion in statistical studies of strength of materials, these results must be considered disturbing, though in none of the studies so far is it clear whether the discrepancy is due to failures of experimental procedure or to something more fundamental.

   We return to this example in Section 9.12.

4.9 OTHER EXAMPLES OF COMPARING SEVERAL SAMPLES

The examples in section 4.8 correspond to the 'one-way layout' in the design of experiments, in which the data consist of several separate samples and the questions of interest concern the relations among the corresponding survivor functions. In this section, two more examples of this kind of problem will be discussed.

The first example is taken from Green (1984). The data consist of strength tests on tungsten carbide alloy containing cobalt. Five different stress rates ranging from 0.1 to 1000 (MNm^{-2} sec^{-1}) were used, and the failure stress noted for 12 specimens at each stress rate (Table 4.3).

Table 4.3 Strength Tests on Tungsten Carbide Alloy

Col. 1: Stress Rate in MNm^{-2} sec^{-1}; Cols. 2–13: Failure Stresses in MNm^{-2}												
0.1	1676	2213	2283	2297	2320	2412	2491	2527	2599	2693	2804	2861
1	1895	1908	2178	2299	2381	2422	2441	2458	2476	2528	2560	2970
10	2271	2357	2458	2536	2705	2783	2790	2827	2837	2875	2887	2899
100	1997	2068	2076	2325	2384	2752	2799	2845	2899	2922	3098	3162
1000	2540	2544	2606	2690	2863	3007	3024	3068	3126	3156	3176	3685

A proportional hazards plot (similar to Figures 4.4 and 4.6) is in Figure 4.7, with the straight lines obtained by fitting a two-parameter Weibull distribution separately to each stress rate. The five lines appear parallel except for the third (stress rate 10) which has almost double the estimated η (17.8 as against 10.1, 9.0, 8.0 and 9.2 for the other four samples). Arguments from fracture mechanics given by Wright, Green and Braiden (1982) (similar to those in section 4.6 above) suggest a model of precisely the form given by (4.7) with $\log \alpha_x$ linear in $\log x$. In this context t is failure stress and x is stress rate. Call this model M_1. In similar fashion to section 4.8, this may be compared with:

Model M_0: common two-parameter Weibull distribution for all levels;

Model M_2: as M_1 but the $\log \eta$ also linearly dependent on $\log x$;

Model M_3: separate two-parameter Weibull distribution at each stress rate.

The techniques for fitting these models are similar to those for the carbon fibres data of section 4.8. In this case models M_0 and M_3 require no more than a simple two-parameter Weibull fit, applied respectively to the combined data from all five stress rates and the five separate sub-samples, while M_1 may be reduced to a two-parameter optimization by the method described in section 4.8.

All four models have been fitted obtaining maximized log likelihoods $l_0 = -441.9$, $l_1 = -430.5$, $l_2 = -430.4$, $l_3 = -425.6$ with respectively 2, 3, 4 and 10 parameters. Likelihood ratio tests support M_1. In this case M_2 is not a particularly satisfactory alternative, since the evidence for a discrepant value of η lies in the middle sample, but a direct test of M_1 against M_3 is also not significant (χ_7^2 statistic 9.8, whereas the 10% point of χ_7^2 is 12.0). The

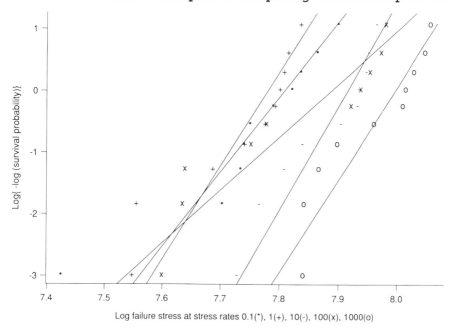

Figure 4.7 Weibull plots for tungsten carbide data, with fitted Weibull lines.

parameter values were given by Green (1984), who also discussed alternative methods of computation.

A second example is given in Table 4.4. The data, taken from Gerstle and Kunz (1983), consist of failure times of Kevlar 49 fibres loaded at four different stress levels. Also shown is the spool number (1–8) from which each fibre was drawn. This information will be used in section 4.10 but for the moment is ignored.

A set of proportional hazards plots is shown in Figure 4.8, with superimposed fitted lines. In this case the four plots are clearly separated, but the data on the highest stress level (29.7) seem to deviate most from the fitted line, and there is again evidence of different Weibull shape parameters (0.52 as against 0.87, 1.01 and 0.95). Note also that these shape parameters are much smaller than those in the previous example, implying vastly greater variability in the data at each stress.

The same four models as above may be considered, in this case producing $l_0 = -864.1$ (2 parameters), $l_1 = -798.3$ (3 parameters), $l_2 = -792.7$ (4 parameters), $l_3 = -790.6$ (8 parameters), which points to M_2 as the preferred model. If we write this in the form of equation (4.8) with $\log \alpha_x = \zeta - \beta \log x$ and $\log \eta_x = \rho - \tau \log x$, then the parameter estimates are $\hat{\zeta} = 88.2$ (standard error 5.1), $\hat{\beta} = 24.5$ (1.6), $\hat{\rho} = 10.5$ (2.9), $\hat{\tau} = -3.3$ (0.9). The fitted values of η

Table 4.4 Kevlar 49 Failure Data

Stress in MPa, Spool Number, Failure time in hours (* if censored)											
Stress	Spool	F-Time	Stress	Spool	F-Time	Stress	Spool	F-Time	Stress	Spool	F-Time
29.7	2	2.2	29.7	5	243.9	27.6	2	694.1	25.5	1	11 487.3
29.7	7	4.0	29.7	4	254.1	27.6	4	876.7	25.5	5	11 727.1
29.7	7	4.0	29.7	1	444.4	27.6	1	930.4	25.5	4	13 501.3
29.7	7	4.6	29.7	8	590.4	27.6	6	1254.9	25.5	1	14 032.0
29.7	7	6.1	29.7	8	638.2	27.6	4	1275.6	25.5	4	29 808.0
29.7	6	6.7	29.7	1	755.2	27.6	4	1536.8	25.5	1	31 008.0
29.7	7	7.9	29.7	1	952.2	27.6	1	1755.5	23.4	7	4 000.0
29.7	5	8.3	29.7	1	1108.2	27.6	8	2046.2	23.4	7	5 376.0
29.7	2	8.5	29.7	4	1148.5	27.6	4	6177.5	23.4	6	7 320.0
29.7	2	9.1	29.7	4	1569.3	25.5	6	225.2	23.4	3	8 616.0
29.7	2	10.2	29.7	4	1750.6	25.5	7	503.6	23.4	5	9 120.0
29.7	3	12.5	29.7	4	1802.1	25.5	3	1087.7	23.4	2	14 400.0
29.7	5	13.3	27.6	3	19.1	25.5	2	1134.3	23.4	6	16 104.0
29.7	7	14.0	27.6	3	24.3	25.5	2	1824.3	23.4	5	20 231.0
29.7	3	14.6	27.6	3	69.8	25.5	2	1920.1	23.4	6	20 233.0
29.7	6	15.0	27.6	2	71.2	25.5	2	2383.0	23.4	5	35 880.0
29.7	3	18.7	27.6	3	136.0	25.5	3	2442.5	23.4	1	41 000.0*
29.7	2	22.1	27.6	2	199.1	25.5	8	2974.6	23.4	1	41 000.0*
29.7	7	45.9	27.6	2	403.7	25.5	2	3708.9	23.4	1	41 000.0*
29.7	2	55.4	27.6	2	432.2	25.5	8	4908.9	23.4	1	41 000.0*
29.7	7	61.2	27.6	1	453.4	25.5	2	5556.0	23.4	4	41 000.0*
29.7	5	87.5	27.6	2	514.1	25.5	6	6271.1	23.4	4	41 000.0*
29.7	8	98.2	27.6	6	514.2	25.5	8	7332.0	23.4	4	41 000.0*
29.7	3	101.0	27.6	6	541.6	25.5	8	7918.7	23.4	4	41 000.0*
29.7	2	111.4	27.6	2	544.9	25.5	6	7996.0	23.4	8	41 000.0*
29.7	6	144.0	27.6	8	554.2	25.5	8	9240.3	23.4	8	41 000.0*
29.7	2	158.7	27.6	1	664.5	25.5	8	9973.0	23.4	8	41 000.0*

are 0.56, 0.71, 0.92, 1.22 for the four stress levels. However there is no reason in this case to suppose that the relation between $\log \eta_x$ and $\log x$ is a linear one.

This discussion is continued in section 4.10. It turns out that some vastly different results are obtained if the spool effect is taken into consideration.

4.10 WEIBULL ANOVA

This section continues the second example of the previous section, using it to introduce a general technique.

One of the main conclusions of Gerstle and Kunz (1983) was that there is significant variation from spool to spool, to such an extent that this must be considered one of the dominant factors in predicting failure time. They reached

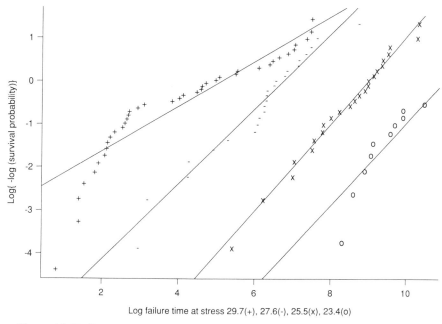

Log failure time at stress 29.7(+), 27.6(-), 25.5(x), 23.4(o)

Figure 4.8 Weibull plots for Kevlar pressure vessel data, with fitted Weibull lines.

this conclusion after applying statistical tests of two broad types: an analysis of variance of log failure times using standard least squares methods, and a non-parametric procedure based on the Kruskal–Wallis test. In this section, a third alternative will be presented: an analysis of variance based on the Weibull distribution with maximum likelihood fitting.

Although the analysis of section 4.9 suggested that the Weibull shape parameter varies from sample to sample, for the reasons explained earlier a model with constant shape parameter is usually preferred, so the first model fitted is (4.7), with

$$\log \alpha_x = \zeta - \beta \log x \qquad (4.17)$$

which produced estimates $\zeta = 86.9$ (standard error 6.0), $\beta = 24.1$ (1.8), $\eta = 0.68$ (0.06). Various probability plots have been produced of portions of the residuals. For these plots the minus sign was taken in (4.12), so we are effectively transforming from minima to maxima and it is the upper tail of the plot which is of greatest interest. Figure 4.9 shows a separate probability plot for each of the eight spools. The idea behind these plots is as follows. Having fitted the model, we could transform the individual observations to uniformity using the probability integral transformation (4.11). Applied to the whole data set, a probability plot would allow us to assess how well the model fits overall.

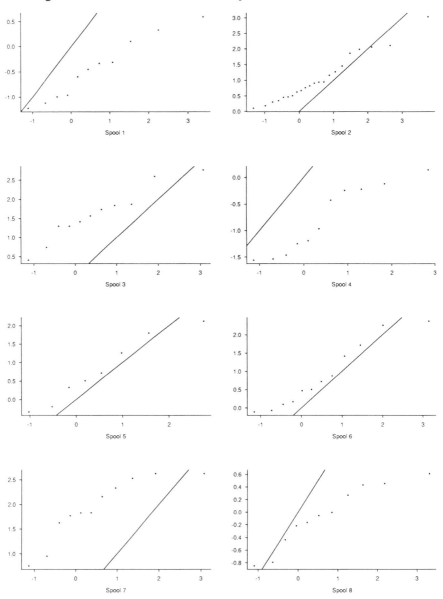

Figure 4.9 Kevlar pressure vessel data: plot of observed *vs.* expected values of residuals, separately for each spool, for the model in which spool effect is not taken into account. If the model were a good fit, the residuals in each plot should cluster round the solid line. The fact that they do not is evidence of a significant spool effect.

In Figure 4.9, the transformation (4.11) followed by (4.12), based on the overall fitted parameters, has been applied separately to the data from each of the eight spools and a probability plot constructed from each. If there is no spool effect, then the z_is from each spool should again look like an independent sample of Gumbel distributed observations. In this case, however, the comparison with the theoretical Gumbel line (the solid line in each of the plots) shows a dramatic spool effect. For spools 1, 4 and 8 the plot lies almost entirely below the fitted straight line, while in the others it is above. The implication is that fibres from spools 1, 4 and 8 are substantially stronger than the others.

The analysis suggested by this is to amend model (4.17) to

$$\log \alpha_{x,j} = \zeta - \beta \log x + \psi_j, \qquad (4.18)$$

where j denotes spool number (1–8). To keep the model identifiable we arbitrarily define $\psi_8 = 0$ so that $\psi_1, \ldots, \psi_7$ represent differences between spools 1–7 and spool 8. This model has been fitted using the earlier fit as starting values for ζ, β and η and taking ψ_1–ψ_7 initially 0.

Results of the fit show a large increase in l ($\chi_7^2 = 111$) with parameter estimates $\hat{\zeta} = 84.1$ (standard error 3.5), $\hat{\beta} = 23.1$ (1.1), $\hat{\eta} = 1.26$ (0.10), $\hat{\psi}_1$–$\hat{\psi}_7$ equal to 0.6, -1.5, -2.3, 1.1, -0.9, -1.1, -2.8, all with standard errors of 0.3 or 0.4. The ordering of spools from strongest to weakest (4, 1, 8, 5, 6, 2, 3, 7) follows closely that concluded by Gerstle and Kunz, though they had the positions of 3 and 7 reversed; interestingly, though, they also commented that in individual experiments spool 7 tended to fail first. Probability plots based on the new model (Figure 4.10) show that most of the earlier discrepancy has now disappeared. There is still some doubt about the upper tail of several plots, notably those for spools, 5, 6 and 8, but it seems unlikely that any further investigation of this will be worthwhile in view of the small sample sizes. It should also be noted that the Weibull shape parameter ($\hat{\eta} = 1/\hat{\sigma}$) has almost doubled, from 0.68 to 1.27, implying that the variability of failure times at each stress level is considerably reduced when spool variability is taken into account.

A final set of plots is given in Figure 4.11. This shows probability plots of the residuals from the final fitted model by stress level. Those for the three highest stress levels show little cause for concern, but the evidence from the lowest stress level (23.4) is that the Weibull shape parameter is different in this case. Thus, even after removing the spool effect, we see the same change in Weibull shape parameter as in section 4.9. However, it should be pointed out that the plot suggests a discrepancy in η rather than α – the scale is correctly reproduced but the shape parameter is wrong. In view of the small sample size at this stress level there seems little prospect of improving matters by further statistical analysis. One possibility which suggests itself is to re-fit the Weibull

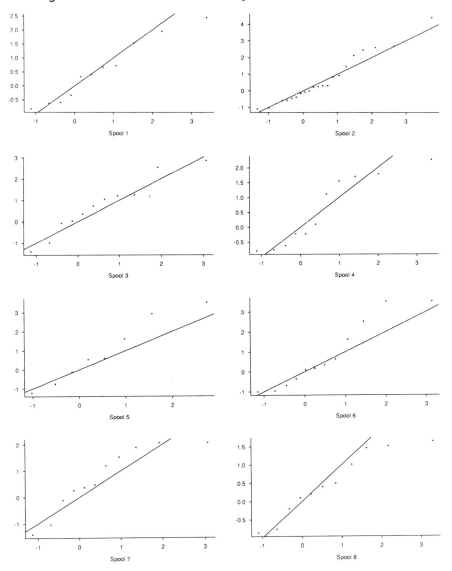

Figure 4.10 As Figure 4.9, but with spool effect now included in the model. In this case there is no evidence of significant discrepancies.

model just to the residuals corresponding to stress 23.4. This, however, assumes that the observed spool variability persists in this stress regime, which must itself be a questionable assumption.

We conclude this section with some discussion of the use of these models to estimate particular quantities of interest. Two kinds of extrapolation are

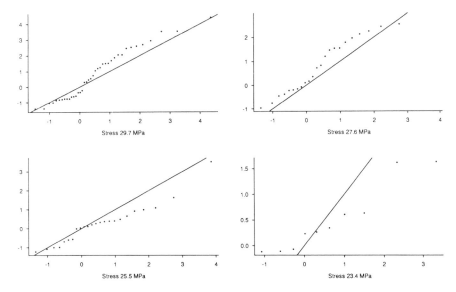

Figure 4.11 As Figure 4.10, but with residuals classified by stress level rather than by spool. There is some suggestion of a different slope at stress 23.4.

of particular interest: (a) extrapolations to very low failure probabilities, and (b) the presumably much longer survival times that would arise under much lower loads than those in the experiment. To illustrate the issues that arise, two specific predictions will be considered: (a) the time to which 99% of fibres will survive at load 23.4, and (b) the median failure time at load 22.5, the latter figure being chosen to represent a load lower than any of those in the experiment but still close enough to provide reasonable extrapolations.

The first model which we use for illustration is model (4.17), with parameter estimates as previously calculated. It has already been shown that this model does not fit the data very well, but it is nevertheless of interest if only to compare the results with those of other models.

Using this model to solve problem (a), we obtain a point estimate (for the time at which 99% of the fibres survive at load 23.4) of 70 hours with 95% confidence interval [22, 225]. The confidence interval is calculated by first expressing log failure time as a function of ζ, β and σ, calculating its standard error by the delta method, then calculating an approximate 95% confidence interval for log failure time as estimate $\pm$ two standard errors, and then exponentiating. A similar calculation for problem (b) gives estimate 88 000 hours with confidence interval from 42 000 and 187 000.

In Table 4.5, the same calculations are reported but this time separately for each of the eight spools, using model (4.18) and the parameter estimates reported earlier. The first point to note is the very wide variation among both point and interval predictions for the eight spools, emphasizing again the importance of taking spool effect into account but also underlining the difficulty of making general statements about the strength of a material, without reference to such local aspects as spool number.

Table 4.5 Point and Interval Estimates of Failure Times for Eight Spools

| | Point estimates for problems (a) and (b) with lower and upper confidence limits | | | | | |
Spool	F-Time (a) (hours)	Lower CL	Upper CL	F-Time (b) ($\times 1000$ hours)	Lower CL	Upper CL
1	3762	1701	8317	263	138	502
2	461	222	957	32.2	19.3	54.0
3	217	95	497	15.2	8.15	28.4
4	6264	2757	14234	438	221	869
5	874	369	2070	61.1	32.0	117
6	709	322	1563	49.6	28.2	87.4
7	131	56	305	9.19	4.72	17.9
8	2108	970	4581	147	79.6	273
All	70	22	225	88	42	187

In the case of problem (a), it is noteworthy that all the point and upper and lower confidence limits for the separate spools are much greater than the corresponding quantities for all spools combined. The explanation is that the model (4.17) leads to an estimate of Weibull shape parameter much smaller than does (4.18). Hence the estimated quantiles are much lower in the lower tail and much higher in the upper tail. The lesson is that ignoring a vital parameter, such as spool effect, may not only be significant in itself but may also lead to bias in estimating the other parameters.

The problem is further complicated by the suggestion that the Weibull shape parameter may be different at different stresses. Gerstle and Kunz also suggested that the shape parameter could be different for different spools! In the light of such contradictory indications about the model, it seems doubtful that any more than very general statements could be made about the correct extrapolations.

To summarize, it is possible to judge the sensitivity of extrapolated quantities to the assumed model and parameters, both by computing standard errors of the estimates and by repeating the analysis under different models. There is enough variability in the values quoted here, but Gerstle and Kunz went on to consider design loads for 99.9999% reliability at a 10-year life span. In view of the difficulties experienced here, there must be doubts about the practicality of that.

4.11* BUFFON'S BEAMS: AN HISTORICAL EXAMPLE OF RELIABILITY DATA

Our final example in this chapter illustrates some more advanced features of the methodology we have been discussing. It is based on an intriguing data set due to Buffon (1775). On an historical note, we may note that both Galileo and Leonardo da Vinci were aware of the size effect on material strength, but Buffon may have been the first to collect and publish a large data set (Table 4.6).

The test specimens consist of beams having cross-sections between 4 and 8 inches, and lengths between 7 and 12 feet. Each beam was supported on two stands, 3 feet high and 7 inches wide, positioned so that it protruded half an

Table 4.6 Buffon's Data

Cols. 1–6: Width of beam in inches, Length of beam in feet, Weight of beam in pounds, Breaking load in pounds, Time in minutes, Sag in twelfths of an inch
(Repeated in Cols. 7–12, 13–18)

1	2	3	4	5	6	7	8	9	10	11	12	13	14	15	16	17	18
4	7	60.0	5 350	29	42	5	18	231.0	3 650	10	98	7	8	201.5	25 950	133	30
4	7	56.0	5 275	22	54	5	20	263.0	3 275	10	106	7	9	227.0	22 800	100	37
4	8	68.0	4 600	15	45	5	20	259.0	3 175	8	120	7	9	225.0	21 900	97	35
4	8	63.0	4 500	13	56	5	22	281.0	2 975	18	135	7	10	254.0	19 650	73	31
4	9	77.0	4 100	14	58	5	24	310.0	2 200	16	132	7	10	252.0	19 300	76	36
4	9	71.0	3 950	12	66	5	24	307.0	2 125	15	162	7	12	302.0	16 800	63	35
4	10	84.0	3 625	15	70	5	28	364.0	1 800	17	216	7	12	301.0	15 550	60	40
4	10	82.0	3 600	15	78	5	28	360.0	1 750	17	264	7	14	351.0	13 600	55	50
4	12	100.0	3 050	0	84	6	7	128.0	19 250	109	*	7	14	351.0	12 850	48	45
4	12	98.0	2 925	0	84	6	7	126.5	18 650	98	*	7	16	406.0	11 100	41	58
5	7	94.0	11 775	58	30	6	8	149.0	15 700	72	28	7	16	403.0	10 900	36	63
5	7	88.5	11 275	53	30	6	8	146.0	15 350	70	29	7	18	454.0	9 450	27	66
5	8	104.0	9 900	40	32	6	9	166.0	13 450	56	30	7	18	450.0	9 400	22	70
5	8	102.0	9 675	39	35	6	9	164.5	12 850	51	34	7	20	505.0	8 550	15	94
5	9	118.0	8 400	28	36	6	10	188.0	11 475	46	36	7	20	500.0	8 000	13	102
5	9	116.0	8 325	28	39	6	10	186.0	11 025	44	42	8	10	331.0	27 800	170	36
5	9	115.0	8 200	26	42	6	12	224.0	9 200	31	48	8	10	331.0	27 700	178	27
5	10	132.0	7 225	21	38	6	12	221.0	9 000	32	49	8	12	397.0	23 900	90	36
5	10	130.0	7 050	20	42	6	14	255.0	7 450	25	54	8	12	395.5	23 000	83	35
5	10	128.5	7 100	18	48	6	14	254.0	7 500	22	50	8	14	461.0	20 050	66	46
5	12	156.0	6 050	30	66	6	16	294.0	6 250	20	66	8	14	459.0	19 500	62	38
5	12	154.0	6 100	0	69	6	16	293.0	6 475	19	70	8	16	528.0	16 800	47	62
5	14	178.0	5 400	21	96	6	18	334.0	5 625	16	89	8	16	524.0	15 950	50	45
5	14	176.0	5 200	18	99	6	18	331.0	5 500	14	102	8	18	594.0	13 500	32	54
5	16	209.0	4 425	17	97	6	20	377.0	5 025	12	114	8	18	593.0	12 900	30	49
5	16	205.0	4 275	15	98	6	20	375.0	4 875	11	106	8	20	664.0	11 775	24	78
5	18	232.0	3 750	11	96	7	8	204.0	26 150	126	33	8	20	660.5	12 200	28	72

inch for every foot of its length. A square iron ring was placed around the middle of the beam and a strong table, on which weights were placed, was suspended from the ring. The stands were fixed to stop their moving outwards during the loading of the weights. Eight men loaded the 200, 150 and 100 pound weights while two men on a scaffold (to avoid being crushed) put the 50 and 25 pound weights on the table. Four men propped the table to prevent it from swinging, and a man with a ruler measured the bending of the plank in its middle corresponding to the load applied. Another man recorded the time and the load applied.

Thus the variables of interest include failure load, failure time (this is the time between the appearance of the first crack and eventual failure) and the sag of the beam in its middle at the time of the first crack. Only the failure load data will be considered here. Galileo's law suggests that this should be of the form

$$P = kab^2/L \qquad (4.19)$$

where P is the breaking load, a is width of beam, b is the thickness or height of beam (in these experiments $a=b$), L is length and k a constant of proportionality.

The obvious starting point is an ordinary least squares regression of log strength against log length and log width. Figure 4.12 shows plots of log strength against log length for the five widths (4, 5, 6, 7, 8 inches). Note that the

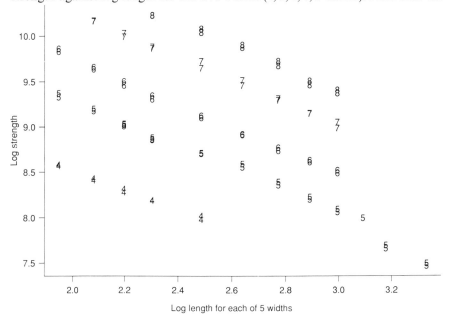

Figure 4.12 Plot of log strength *vs.* log length for Buffon data, for each of 5 widths (4–8 inches).

data set consists of pairs of experiments at the same width and length, and consequently there are many points nearly or exactly superimposed in the plot. The plot confirms a linear relationship but with slope about -1.25 rather than -1 as suggested by (4.19). Also the spacing between widths is not even. A least squares fit of the model

$$\log P = \beta_0 + \beta_1 \log a + \beta_2 \log L + y, \qquad (4.20)$$

where y is assumed Normally distributed, produces parameter estimates $\beta_1 = 2.98$ (standard error 0.02) and $\beta_2 = -1.25$ (0.01). This is consistent with what we expected for β_1, but not for β_2.

By analogy with the models studied in earlier sections, however, it would be more reasonable to assume that the random component y has either a Gumbel or a generalized extreme value distribution, the former corresponding to a two-parameter Weibull distribution for the original failure loads. As a test of which is more appropriate, probability plots of the residuals from the Normal model just fitted are shown in Figures 4.13 and 4.14. Figure 4.13 shows a Weibull plot based on (4.12) with the minus sign. The marked curvature at the top end indicates a sharp departure from the Weibull distribution in the lower

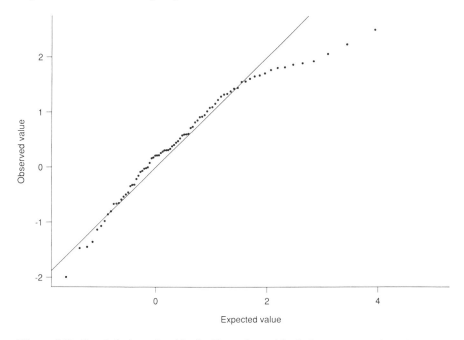

Figure 4.13 Gumbel plot of residuals. First, the residuals from a normal model of log strength *vs.* log width and log length were calculated. Then, a Gumbel distribution was fitted to those residuals, and the residuals from that fit are plotted here.

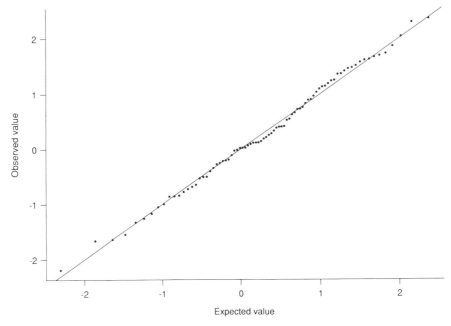

Figure 4.14 As Figure 4.13, but a generalized extreme value, instead of Gumbel, distribution was fitted to the Normal residuals.

tail of the original data. In contrast Figure 4.14, which shows an observed versus expected values plot after fitting a generalized extreme value distribution to the Normal residuals, shows good fit by this distribution. These results suggest that we should re-fit the model (4.20) by maximum likelihood, assuming y to be generalized extreme value – equivalent to the model (4.10) in which σ and γ are constants and $\mu = \beta_0 + \beta_1 \log a + \beta_2 \log L$. The most important parameters are $\beta_1 = 3.01$ (standard error 0.03), $\beta_2 = -1.26$ (0.02), $\gamma = -0.48$ (0.07). Thus the new fit is very little different as far as β_1 and β_2 are concerned, but does provide highly significant evidence in favour of the generalized extreme value distribution as against the Gumbel distribution.

An additional issue which arises here is the closeness of γ to the value -0.5, which as explained in section 4.7 is the value at which classical maximum likelihood theory breaks down. In view of this, it seems worthwhile to try an alternative method of fitting. Smith (1989) proposed a method of fitting a non-regular regression model by first fitting a hyperplane under the data points (the algorithm minimizes the sum of residuals subject to the constraint that all residuals be non-negative) and then fitting the remaining parameters by maximum likelihood fitting to the positive residuals. The full details are beyond the scope of the present discussion, but the procedure leads to estimates $\beta_1 = 3.08$, $\beta_2 = -1.24$, $\gamma = 0.54$ (standard error 0.05), very little different from the previous estimates, but providing some evidence that the non-regularity is not seriously distorting the estimates.

If the analysis ended at this point, our conclusion would be that the Gumbel model did not provide a very good fit and the generalized extreme value model was superior. In fact, this conclusion is incorrect and illustrates once again the need for taking into account external sources of variability. The data have been presented in the order given in Buffon's book, and there is no indication from Buffon's description that they represent the actual time order in which the experiments were conducted. Nevertheless, a plot of residuals against time (Figure 4.15) shows a strong time-dependent effect, the reasons for which are something of a mystery. It might appear that some form of sinusoidal regression would be appropriate to describe this figure, but after some trial and error it was found that an autoregressive model of order 3 (combined with the regression on $\log \alpha$ and $\log L$) fits the data very well. A Gumbel plot from this model is in Figure 4.16, and again appears close to a straight line. However, there is no substantial change in the estimates of β_1 and β_2 so the failure to fit Galileo's law still has to be explained.

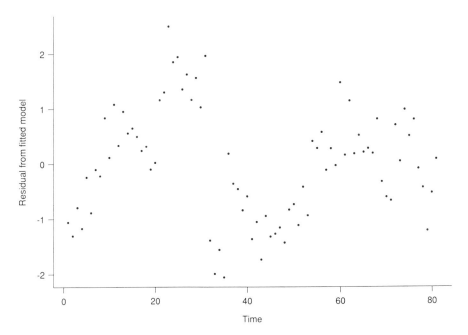

Figure 4.15 Buffon data: the residuals from the Normal model (as for Figure 4.13) plotted against 'time' as represented by the order of the data. There appears to be long-term cyclicity in the data, though closer analysis reveals that this is very well explained by an autoregressive model of order 3.

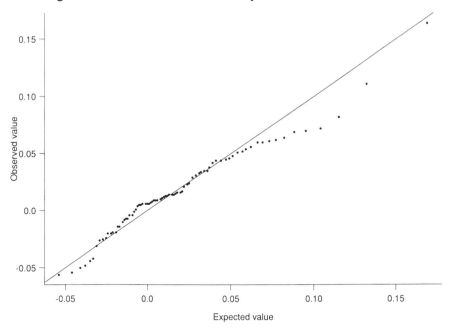

Figure 4.16 Gumbel plot of residuals from a model in which log strength was regressed on log width and log length, with errors assumed to follow an autoregressive model of order 3. In contrast with Figure 4.13, in which the errors were assumed independent, the fit now appears acceptable.

This example was first brought to our attention by Stephen M. Stigler of the University of Chicago; it is a pleasure to acknowledge a number of comments about the data from Professors Stigler and David L. Wallace.

4.12 CONCLUDING REMARKS

This chapter began by reviewing some standard survival data models such as accelerated life, proportional hazards and proportional odds. In the reliability context, arguments from physics sometimes suggest the accelerated life model as the most appropriate, though the evidence of this from experimental data is often equivocal. Several of our examples have featured attempts to incorporate physical considerations into the model, though with mixed results. Some authors have advocated an approach in which the model is determined entirely by the data. However, the difficulty with such an approach is that it leaves one in difficulties as soon as any form of extrapolation is attempted. At the same time, it would be very bad science to fit the model suggested by physics without also attempting to check up on the fit of the model.

A final comment, which has not been emphasized at all but which may be very important, is the need for a properly randomized experimental design.

Indeed, several examples have suggested sources of experimental variability which are additional to natural variability in the materials under study, and it is possible that a better experimental design would have eliminated these.

The examples illustrate the wide range of models that can be fitted by numerical maximum likelihood techniques, and some of the diagnostic techniques that can be used to check the fit of such models. As further advances are made in statistical methodology, we may surely expect them to be used to the benefit of improved reliability prediction.

5

Proportional hazards modelling

5.1 INTRODUCTION

In this chapter we discuss the analysis of reliability data using the proportional hazards model introduced in Chapter 4. Since Cox's (1972) pioneering paper on the subject, the vast majority of reported uses of proportional hazards methods have been for the analysis of survival data in the medical field. The semiparametric hazards-based approach however is beginning to have more of an impact in the reliability field, and is likely to play a more prominent rôle in the future, since no assumptions regarding the form of the baseline hazard function are necessary for the analysis. Much of the practical reliability literature in this area is contained in internal reports or conference proceedings; Wightman and Bendell (1985) provide a number of such references which include applications to marine gas turbines, motorettes, nuclear reactors, aircraft engines and brake discs on high-speed trains. See also Ansell and Ansell (1987) and Drury *et al.* (1987). Ascher (1982) discusses the application of proportional hazards modelling to repairable systems reliability.

The proportional hazards (PH) model introduced in Chapter 4 models the effects on the hazard function of certain explanatory variables or factors associated with the lifetime of equipment. The linear PH model assumes that, under conditions given by a vector $x = (x_1, \ldots, x_n)^T$ of concomitant variables, the hazard function is of the form

$$h(t; x) = h_0(t)e^{x^T \beta} \tag{5.1}$$

where β is a vector of unknown parameters $\beta_1, \ldots, \beta_p$, and h_0 is some baseline hazard function. Thus the combined effect of the x-variables is simply to scale the hazard function up or down.

The statistical analysis depends on whether or not we assume a particular functional form for the baseline hazard function $h_0(t)$. In the former case, we can write the baseline cumulative hazard function H_0 in the form $G(t; \theta)$ where G is some known function and θ an unknown vector of parameters. One important case is $G(t; \theta) = \alpha G(t)^{\eta}$ where G is a specified non-decreasing function of t. The survivor function $S(t; x)$ under conditions given by x is $S(t; x) = \exp\{-H(t; x)\}$ where $H(t; x) = \int_0^t h(s; x)\,ds$ is the corresponding cumulative hazard function. Therefore we have

$$S(t; x) = \exp\{-H_0(t)e^{x^T\beta}\}$$
$$= \exp\{-\alpha G(t)^{\eta}e^{x^T\beta}\}$$
$$= \exp\{-\exp(\eta y + \log\alpha + x^T\beta)\}$$

where $y = \log G(t)$. It follows that the PH model can be written in the form

$$y = -\eta^{-1}\log\alpha - x^T(\eta^{-1}\beta) + \eta^{-1}z \tag{5.2}$$

where the random variable z has the standard Gumbel distribution. Thus in this case the PH model is equivalent to a log-linear Weibull model for the transformed times $G(t_i)$, the analysis of which has been discussed in Chapter 4.

In the case where h_0 is unspecified, one simple approach to statistical analysis is to assume a particular initial form for the baseline cumulative hazard function H_0, such as αt^{η}, fit a log-linear model as described above, and use residual analysis to identify an appropriate transformation $G(t)$ of the data. This procedure may then be repeated until a satisfactory fit is obtained. In the following sections, we shall describe a direct method of estimating the regression coefficients β from a 'likelihood function' which does not involve the unknown hazard function h_0.

5.2 ANALYSIS OF THE SEMIPARAMETRIC PH MODEL

We suppose that the function h_0 in (5.1) is completely unspecified. In this section we discuss the meaning and application of a form of likelihood for β which does not involve the unknown function h_0. The ideas presented here are developed much more fully in Kalbfleisch and Prentice (1980).

Suppose then that a random sample of n items yields r distinct lifetimes and $n - r$ censoring times. Let the ordered lifetimes be $t_{(1)}, \ldots, t_{(r)}$ and let R_i be the set of items with $t \geq t_{(i)}$, where t may be either an observed lifetime or a censored time. Here R_i is the **risk set** at time $t_{(i)}$; that is, those items which were at risk of failing just prior to $t_{(i)}$. Estimation of β may then be based on the 'likelihood function'

$$L(\beta) = \prod_{i=1}^{r} \frac{\exp(x_{(i)}^T\beta)}{\left[\sum_{l \in R_i} \exp(x_l^T\beta)\right]} \tag{5.3}$$

where $x_{(i)}$ is the vector of regressor variables associated with the unit observed to fail at time $t_{(i)}$.

There are several alternative ways of justifying the use of the above likelihood function. The simplest argument is due to Cox (1972). Consider the set R_i of items at risk immediately prior to $t_{(i)}$. The conditional probability that the item corresponding to $x_{(i)}$ is the next to fail, given the failure times $t_{(1)}, \ldots, t_{(i)}$, is

$$\frac{h(t_{(i)}; x_{(i)})}{\sum_{l \in R_i} h(t_{(i)}; x_l)} = \frac{\exp(x_i^T \beta)}{\sum_{l \in R_i} \exp(x_l^T \beta)}$$

The likelihood (5.3) is now formed by taking the product of these terms over all failure times $t_{(i)}$. When there are no censored observations, the expression (5.3) then gives the joint probability of the sample, conditional on the observed failure times $t_{(1)}, \ldots, t_{(r)}$, and is referred to as a **conditional likelihood**. In the presence of censored observations, (5.3) is a form of **partial likelihood** (Cox 1975). One now argues that, in the absence of knowledge of $h_0(t)$, the observed failure times themselves supply little or no information on the regression coefficients β so that (5.3) is an appropriate likelihood for inference about β.

An alternative justification for the use of (5.3) is due to Kalbfleisch and Prentice (1973). In the case of uncensored, or Type II censored samples, (5.3) is the **marginal likelihood** of β based only on the **ranks** of the observations. That is, only the *ordering* of the items with respect to their failure times is considered, and the actual failure times ignored. In this derivation, the idea that the lifetimes themselves provide 'no information on β' can be expressed in a more formal way. The argument may also be extended to cover the general case of censored observations. We also mention Bayesian interpretations of (5.3) in Chapter 6.

Generally, then, there is fairly broad consensus that (5.3) is an appropriate likelihood to use when there is very little information on the form of $h_0(t)$. Treating (5.3) as an ordinary likelihood, the general methods of maximum likelihood (ML) estimation discussed in Chapter 3 may be used to estimate β, and approximate standard errors may be obtained from the matrix of second-order derivatives of $\log L(\beta)$ evaluated at the ML estimator $\hat{\beta}$. Models may be compared by means of their respective maximized log-likelihoods, as in Chapter 3.

Computation

For the computation of (5.3), calculate the values $\lambda_l = \exp(x_l^T \beta)$ and accumulate these from the largest life/censoring time to the smallest, giving values $c_i = \sum_{l \in R_i} \lambda_l$, $i = 1, \ldots, r$. The likelihood $L(\beta)$ is then the product of the terms $\lambda_{(i)}/c_i$ for $i = 1, \ldots, r$, where $\lambda_{(i)} = \exp(x_{(i)}^T \beta)$. The log-likelihood function $l(\beta)$ and its derivatives are given by

$$l(\beta) = \sum_{i=1}^{r} (x_{(i)}^T \beta - \log c_i)$$

$$l'(\beta) = \sum_{i=1}^{r} (x_{(i)} - v_i)$$

$$l''(\beta) = -\sum_{i=1}^{r} (A_i - v_i v_i^T)$$

where

$$v_i = c_i^{-1} \sum_{l \in R_i} \lambda_l x_l$$

and

$$A_i = c_i^{-1} \sum_{l \in R_i} \lambda_l x_l x_l^T.$$

A numerical maximization procedure, or a procedure such as Newton–Raphson to solve the likelihood equation $l'(\beta) = 0$, is then used to obtain $\hat{\beta}$. Some Fortran programs are given in Kalbfleisch and Prentice (1980), and routines exist within the BMDP and SAS packages. For **small** data sets, it is possible to use GLIM, as described by Whitehead (1980). Generally, care should be taken when there are large values of covariates, otherwise overflow may occur in $\exp(x^T \beta)$; suitable scaling of the x-variables should remedy this.

Tied observations
One complication in the above analysis occurs when there are *ties* in the failure-time data, which may occur due to rounding errors. Expressions for the conditional and marginal likelihoods in such cases are quite complex. In both cases, however, a reasonable approximation is often afforded by the expression

$$L(\beta) = \prod_{i=1}^{r} \frac{\exp(s_i^T \beta)}{\left[\sum_{l \in R_i} \exp(x_l^T \beta) \right]^{d_i}} \tag{5.4}$$

where d_i is the number of items failing at time $t_{(i)}$, and s_i is the sum of the covariates of items observed to fail at $t_{(i)}$. This approximation is satisfactory provided the ratios d_i/n_i are small, where n_i is the number of items at risk at time $t_{(i)}$. If these ratios are sizeable however, then it may be better to fit a discrete version of the PH model.

Time-dependent covariates

In reliability applications it is very often appropriate to relate the current risk of failure to a covariate which is *time-dependent*, such as an estimate of the current wear or damage sustained by the component. To deal with this it is necessary to replace x by $x(t)$ in the PH model (5.1). The method described above may be adapted when one or more or the covariates is time-dependent. One can distinguish several different types of time-dependent variable. The primary distinction is between an external and an internal covariate. An **external covariate** is one which is not directly involved with the failure mechanism, such as an applied stress, while an **internal covariate** is a time measurement generated by the item itself.

Cox (1975) proposed the use of the partial likelihood

$$L(\beta) = \prod_{i=1}^{r} \frac{\exp(\{x_{(i)}(t_{(i)})\}^T \beta)}{\left[\sum_{l \in R_i} \exp(\{x_l(t_{(i)})\}^T \beta) \right]} \tag{5.5}$$

in such cases, when there are no tied values. If ties are present, a suitable approximation is (5.4) with x replaced by $x(t_{(i)})$. Computations relating to (5.5) turn out to be rather more time-consuming than those for (5.3).

5.3 ESTIMATION OF THE SURVIVOR AND HAZARD FUNCTIONS

Once the regression coefficients have been estimated by maximization of (5.3), or one of its modifications, it is possible to obtain a non-parametric estimate of the baseline survivor function. The following formula is a generalization of the PL (product limit) estimator described in Chapter 2, and is appropriate when there are no ties.

$$\hat{S}_0(t) = \prod_{j:t_{(j)} < t} \hat{\alpha}_j \tag{5.6}$$

where

$$\alpha_j^{\lambda_j} = 1 - \left(\lambda_j \bigg/ \sum_{l \in R_j} \lambda_l \right) \tag{5.7}$$

and $\lambda_j = \exp(x_j^T \beta)$. The $\hat{\alpha}_j$ values required in (5.6) are obtained by substituting $\hat{\beta}$ for β in (5.7). The case of ties is a little more complicated and we refer the reader to Kalbfleisch and Prentice (1980) or Lawless (1982) for details. As with the PL estimator, if the last observation is a censored value C, then $\hat{S}_0(t)$ is undefined past C.

A plot of the empirical survivor function $\hat{S}_0(t)$ may suggest a suitable parametric form for $S_0(t)$. Alternatively, the methods of Chapter 3 may be used here on replacing the PL estimator by $\hat{S}_0(t)$ defined above. It is suggested that the plotting positions p_i be taken as $p_i = 1 - \frac{1}{2}[\hat{S}_0(t_{i-1}+0) + \hat{S}_0(t_i+0)]$. An estimate of the baseline cumulative hazard function is $\hat{H}_0(t) = -\log \hat{S}_0(t)$, or some modification of this.

5.4 MODEL CHECKING

The empirical cumulative hazard function (EHF) $\hat{H}_0(t)$ may be used to check the PH assumption. In order to check the proportionality assumption with respect to a particular factor, **stratify** the data according to the level of the factor and fit a model of the form

$$h_j(t; x) = h_{0j}(t)e^{x^T \beta} \tag{5.8}$$

where j is the factor level. Then, writing $\hat{H}_j(t)$ for the EHF in the jth stratum, the functions $\log \hat{H}_j(t)$ should appear roughly vertically parallel when plotted against t. A similar method may be used for checking the PH assumption with respect to a variable x by grouping neighbouring values of x if necessary, and fitting separate baseline hazard functions as above. For details see Lawless (1982). Some formal goodness-of-fit procedures are given in Schoenfeld (1980), Andersen (1982), Moreau *et al.* (1985) and Gill and Schumacher (1987).

For further model checking, a **residual** may be defined for an uncensored observation by

$$e = \hat{H}_0(t) \, e^{x^T \beta}$$

where t is the observed lifetime and x the corresponding vector of regressor variables. It is also possible to define a residual at a *censored* value by

$$e = \hat{H}_0(t) \, e^{x^T \beta} + 1$$

where now t is the censoring point. If the PH model is adequate, the set of residuals should appear to behave like a sample from the unit exponential distribution. Equivalently, the values $r = \log e$ should appear to behave approximately like a random sample from the Gumbel distribution (see Chapter 2). We note that these residuals can exhibit some peculiarities, and their interpretation is sometimes unclear. An example is given by Ansell (1987) where residual plots appear acceptable when the PH model is inadequate. Such plots should not therefore be used in isolation, but in conjunction with other methods of model checking. Finally, plotting the r-values against other variables x not included in the current model will indicate the possible need for their inclusion in a new model.

The form of $l'(\beta)$ given in section 5.2 suggests an alternative definition of a residual. Schoenfeld (1982) defines the residual corresponding to the ith lifetime to be the $p \times 1$ vector $x_{(i)} - \hat{v}_i$.

These residuals may also be used to assess the adequacy of the PH assumption. An application to the reliability of turbochargers on railway carriages is described in Pettitt and Bin Daud (1990). Here the failure response t is distance travelled, rather than time. Suitably smoothed residual plots reveal a non-PH effect of the time of installation of the turbocharger unit.

5.5 NUMERICAL EXAMPLES

In this section we use proportional hazards methods to investigate two sets of failure data, both of which have been discussed in previous chapters.

The first of these is the set of tungsten carbide stress fatigue data in Green (1984), reproduced in Table 4.3. The background to these data is given in Chapter 4. Figure 4.7 indicates that failure stress increases with stress rate, and so the hazard rate decreases with stress rate. A PH model for these data therefore takes the form

$$h(t; x) = h_0(t)e^{m(x)}$$

where here t is a failure stress, $m(x)$ is some specified decreasing parametric function of stress rate x, and h_0 the baseline hazard function. A suitable tentative form for $m(x)$ is $m(x) = -\beta \log x$, which gives $h(t, x) = h_0(t)x^{-\beta}$. Note that this form implies that the hazard rate tends to zero as the stress rate tends to infinity, and vice versa; this is also the form of hazard function which arises from model M_2 defined in Chapter 4, section 4.9. The ML estimate of β obtained from the likelihood (5.3) is found to be $\hat{\beta} = 0.2069$, and $l''(\hat{\beta}) = -478.81$, giving an estimated standard error of $1/\sqrt{478.81} = 0.0457$. There is therefore very strong evidence here of a positive value for β, as anticipated. Figure 5.1 shows the plot of $\log \hat{H}_0(t)$, obtained as described in section 5.3, against $\log t$. The broad linearity of the plot here supports a log-linear Weibull model for these data, as was assumed in section 4.9 of Chapter 4.

In order to check the proportional hazards assumption, the data were stratified into the five stress rate groups. The empirical hazard function in each group was then obtained as described in section 5.4, and plotted against $\log t$, shown in Figure 5.2. Note that the multiplicative factor $x^{-\beta}$ cannot be included in the stratified model, since stress rate is completely confounded with the stratum chosen here. This is equivalent to the probability plots in Figure 4.7, with the axes inverted. The five hazard functions appear to be vertically parallel, apart from that corresponding to a stress rate of $10\,\text{MNm}^{-2}\,\text{sec}^{-1}$, which is much steeper. Generally, note that such plots based on small amounts

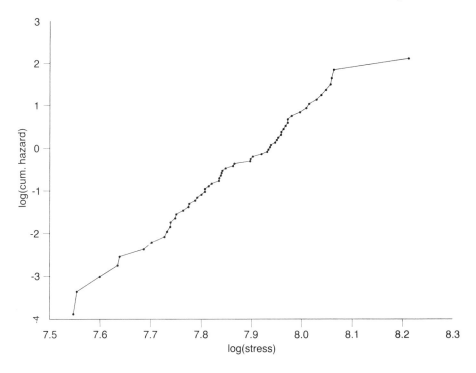

Figure 5.1 Log cumulative hazard plot for tungsten carbide data.

of data are subject to considerable sampling variation, particularly at the extremes.

Since there are replicate observations at each stress rate, it is possible to test for linearity in $\log x$ of the function $m(x)$. In order to illustrate this, we assume that the PH model is adequate, and fit a PH model with separate constants at each stress rate. That is, we fit a model with

$$m(x) = \sum_{j=1}^{5} \alpha_j z_j(x)$$

where $z_j(x) = 1$ if x is at stress level j, and $z_j(x) = 0$ otherwise. We note that, as it stands, this model is not estimable, since an arbitrary constant γ can always be added to each α_j without affecting the model. As is the case with standard regression methods, it is necessary to reduce the dimension of the parameter space by one. This may be achieved by imposing a constraint on the parameters, such as $\sum_{j=1}^{5} \alpha_j = 0$, or simply by setting one of the α_js to zero. Whichever method is used, when the resulting four parameter PH model is fitted, the maximized log-likelihood is found to be -177.00. This may be

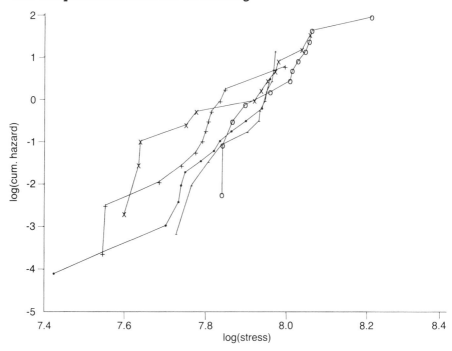

Figure 5.2 Log cumulative hazard plot for tungsten carbide data: stratified model. Stress rates 0.1(.), 1(+), 10(−), 100(×), 1000(0).

compared with the corresponding value of −178.19 under the linear model, giving 2(−177.00 + 178.19) = 2.38 as the value of a $\chi^2(3)$ variable under the hypothesis of linearity. Comparison with the upper 5% point of 7.81 indicates no evidence to reject the hypothesis.

As a second example, we reconsider the Kevlar 49 failure data given in Table 4.4, taken from Gerstle and Kunz (1983). These data are described in Chapter 4, where log-linear Weibull models were fitted. In particular, there was found to be strong evidence in the data of a spool effect, as was discovered by Gerstle and Kunz. An S–N plot indicates increasing hazard with increasing stress, and a plausible PH model incorporating spool-to-spool variation for these data takes the form

$$h(t) = h_0(t) \exp\left(\gamma \log S + \sum_{j=1}^{8} \alpha_j z_j\right) \tag{5.9}$$

where S = stress level/27.0 and $z_j, j = 1, \ldots, 8$ are dummy variables corresponding to the different spools: $z_j = 1$ if spool j was used to wind the vessel, otherwise

$z_j = 0$. As discussed in the previous example, the number of spool-parameters must be reduced by one. For consistency with the treatment in Chapter 4, section 4.10 we set α_8 to zero.

For a speedy iteration in this eight-parameter model, it is advisable to have good initial estimates of the parameters. An obvious method is initially to fit a log-linear Weibull model for the data, either by ML or simply by using least squares, as described in section 4.7. Initial estimates of γ and α may then be deduced from this fit; the connections may be obtained from (5.2). In the present example this approach led to convergence in just four iterations, using a Newton–Raphson procedure. The resulting parameter estimates are given in Table 5.1. The ordering of spool numbers from strongest to weakest is therefore (4, 1, 8, 5, 6, 2, 3, 7), which is the same as that obtained in section 4.10 based on a Weibull ANOVA. The first diagonal element of $[-l''(\hat{\gamma}, \hat{\alpha})]^{-1}$ is 9.594, giving an estimated standard error of 3.097 for $\hat{\gamma}$, so there is very strong evidence that γ is positive. The log-likelihoods under this model, and under the reduced model with all the α_j set to zero, are -261.775 and -320.119 respectively, giving $2(-261.775 + 320.119) = 116.69$ as a possible value of a $\chi^2(7)$ variable under the hypothesis of no spool effect. Thus there is clear evidence here of real spool-to-spool variability, as was concluded in section 4.10.

Table 5.1 Parameter estimates for the model (5.9)

Parameter	γ	α_1	α_2	α_3	α_4	α_5	α_6	α_7
Estimate	32.586	-0.843	2.175	2.878	-1.500	0.962	1.485	3.457

Next, the PH assumption with respect to different spools was examined, which was achieved by stratifying the data by spool number. The stratified model (5.8) with the single regressor log S was then fitted. The corresponding hazard plots are shown in Figure 5.3. In attempting to interpret these plots, it must be borne in mind that they are based on rather small sample sizes (only 8 observations on spool 5, for example), and are particularly variable at the extremes. These plots tend to support the proportionality hypothesis, although there may be some departure at the lower end. It should also be remembered that if the primary interest is in prediction of long term failure, that is failure at relatively low stress levels (c.f. Gerstle and Kunz, 1983), then any deviation at the lower end will probably be of little consequence.

Finally, Figure 5.4 shows the log-hazard plot under the PH model. There does appear to be some departure from linearity here although, again, the high degree of variability at the lower end of the plot should be borne in mind. In fact, an increase in slope in $\log H_0(t)$ would imply some lack of homogeneity of variance in a log-linear Weibull fit, as suggested by the analysis of section 4.9. If one accepts a Weibull model for these data, then the underlying slope of the relationship in Figure 5.4 will give the Weibull shape parameter. A line

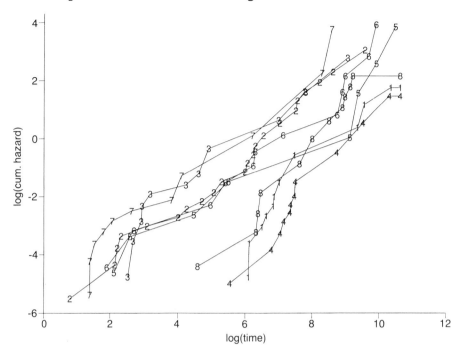

Figure 5.3 Log cumulative hazard plot for Kevlar 49 failure data: stratified model. Spool numbers 1–8.

fitted by eye gives a slope of about 1.29, which is reasonably close to the maximum likelihood estimate of 1.27 obtained in section 4.10. Under a Weibull model, the relationship between the parametrization used here and that of section 4.10 is $\beta = \eta^{-1}\gamma, \psi_j = -\eta^{-1}\alpha_j$. Using the graphical estimate of 1.29 for η, we obtain the estimate 25.3 for β (maximum likelihood estimate 23.1), and ψ_j estimates close to those given in section 4.10.

A more thorough analysis would include an investigation of a possible spool × stress interaction effect. In order to check this a further seven regressors of the form $z_j \log S$ need to be included in the model. Full details are not given here; it turns out that the log-likelihood under the full interaction model is -259.241, yielding a $\chi^2(7)$ value of $2(-259.241 + 261.775) = 5.07$, which is not statistically significant.

In order to illustrate the use of the PH fit for the prediction of long-term failure, one could linearly extrapolate from the latter portion of the log-hazard plot shown in Figure 5.4. A straight line fitted by eye gives

$$\log H_0(t) = -11.19 + 1.533 \log t.$$

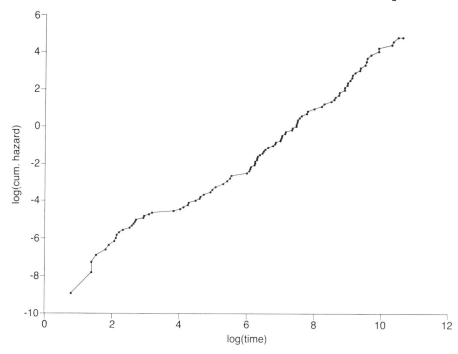

Figure 5.4 Log cumulative hazard plot for Kevlar 49 failure data.

Note that this corresponds to fitting a Weibull model to the upper tail of the baseline failure time distribution, and yields a higher value for η. In conjunction with the PH model parameter estimates in Table 5.1, this yields the long-term failure estimates

$$\hat{H}_j(T) = 15.296 T^{1.533} S^{32.586} e^{\hat{\alpha}_j}$$

where T is time in years ($= t/8766$) and $H_j(T)$ is the cumulative hazard function for spool j. For example, for the predicted median failure time $\tilde{T}$ at 23.4 MPa for spool number 4, set $\hat{H}_j(t) = \log 2$ to give $\tilde{T} = 7.4$ years. The pressure required for a ten-year life with 0.999999 reliability is obtained on setting $T = 10$, $\hat{H}_j(10) = -\log 0.999999$, and gives the estimate 15.273 MPa. This converts to 2214.6 psi, which is lower than the estimate of 2856 psi obtained in Gerstle and Kunz, who assume lognormal failure time distributions. According to (5.9), this latter figure would give a ten-year reliability of 0.996. Similarly, design-allowable life values with any specified level of reliability may be obtained.

We conclude this chapter with the comment that predictions at high levels of reliability, such as those obtained above must be used with extreme caution, as they involve extrapolation of the fitted PH model, as well as that of the non-parametric baseline hazard function. Further analysis and sensitivity studies would be in order here; for example, investigate the sensitivity of those predictions of interest to departures from the assumed power law $H(t) \propto S^{\gamma}$, for which there is some evidence in the data. As is the case with other statistical procedures, the PH method should never be applied blindly. Careful checking of underlying assumptions is always in order, along with appropriate remedial action if necessary.

6

The Bayesian approach

6.1 INTRODUCTION

In this chapter the Bayesian approach to the analysis of reliability data is introduced. Up to this point, statistical inference has been discussed from the classical, or frequentist, point of view. That is, estimators and test statistics are assessed by criteria relating to their performance in repeated sampling. In the Bayesian approach, direct probability statements are made about unknown quantities, conditional on the observed data. This necessitates the introduction of prior beliefs into the inference process.

Bayesian methodology is being used increasingly in the field of reliability data analysis. Apart from the adoption of a Bayesian point of view on philosophical grounds, a major reason for this is that it is possible to incorporate all relevant available information into an analysis in a straightforward manner. We present here only an overview of the main features of the Bayesian approach, with particular emphasis on reliability data analysis. In section 6.2 we give a brief descriptive review of the philosophical basis and practical application of the Bayesian approach, while in section 6.3 we introduce the principal ideas by means of a simple example. Some further general issues are discussed in section 6.4, with particular regard to reliability analysis. Section 6.5 is an introduction to Bayesian decision analysis, a topic which is likely to play an increasingly important rôle in the reliability field. Finally, in section 6.6 we discuss the Bayesian analysis of parametric and semiparametric regression models for reliability data.

6.2 A REVIEW OF THE BAYESIAN APPROACH TO STATISTICS

It can be argued that performance in repeated sampling is not an appropriate basis for drawing inferences from observational data. Long-run performance indicators, such as the standard error of an estimator, may not fairly reflect the precision of an estimate obtained from the actual observed sample. In the Bayesian approach to statistical inference probability statements about the unknown parameters are made conditionally on the observed data. In order to do this it is necessary to accord the parameter θ the status of a random variable. The key feature of the Bayesian approach is that it is necessary to specify the probability distribution of θ prior to observing the sample data; this distribution is called the **prior distribution** of θ. In certain cases, a parameter may have a natural frequency distribution. For example, suppose it is required to estimate the unknown proportion θ of defectives in a particular batch of components. The parameter θ will normally vary from batch to batch, and this frequency distribution may be approximately known from historical data. Often however in Bayesian statistics, probabilities relating to unknown parameters are interpreted as degrees of belief. That is, the prior distribution of θ is a subjective probability distribution. Once the prior distribution is specified, and the data observed, an application of Bayes Theorem yields the posterior distribution of θ, which specifies beliefs about θ, given the sample data. Thus Bayes Theorem is the mechanism by which one's prior beliefs are revised in the light of the sample information.

There is still much discussion regarding the foundations of statistical inference. We shall not explore such matters further here; the interested reader can refer to the book by Barnett (1982) and the many references cited therein. At the very least, we believe that a study of the Bayesian approach to statistics, as a complement to other approaches, is valuable for a proper understanding of the problems of statistical inference, and often serves to highlight common misconceptions held by many users of statistical methods.

From the point of view of reliability data analysis, the adoption of a Bayesian approach has a number of practical advantages. In many reliability problems, the data are simply inadequate for predictions to be made with any reasonable degree of confidence. Thus practical judgements must inevitably be guided by such things as engineering experience, information extracted from reliability data banks and general handbook data. The Bayesian approach provides a means of quantifying such prior information and a logical framework for combining this with sample data in order to arrive at the final inference. In practice the statistical analysis may be reported for a number of alternative sensible prior specifications. Of particular interest in reliability data analysis is the ease with which the Bayesian approach is able to handle censored data. At a more foundational level, the case for the subjectivist approach in reliability is argued strongly in Singpurwalla (1988).

In cases where only rather vague prior knowledge is available, it often does not matter too much exactly how this weak information is quantified. The reason for this is that, when the sample size is sufficiently large, the amount of information available from the sample outweighs the prior information, so that the exact form of prior distribution used is unimportant. In fact, this phenomenon can occur even when individuals hold initially quite different beliefs about the parameters; given sufficient sample information, such individuals will often end up drawing similar inferences from the data. This is sometimes referred to as the **principle of stable estimation**. Related to this, there are large-sample (asymptotic) results analogous to large-sample ML results in classical statistics. The effect of these results is that, in sufficiently large samples, a Bayesian analysis and a frequentist analysis may lead to similar conclusions. In small samples, or in samples containing many censored observations, however, the results may be quite different. This point is of particular relevance in reliability data analysis, since it is quite common for the data to be rather sparse and/or heavily censored.

Finally we mention the natural rôle of Bayesian statistics in a decision analysis. Obtaining the posterior distribution of a parameter solves the inference problem, but very often in reliability analysis the purpose of the study is to support a decision-making process. We return to this topic in section 6.5.

6.3 ELEMENTS OF BAYESIAN STATISTICS

In this section we describe the general application of the Bayesian approach to statistical problems. The main features will be introduced by means of a specific example. We begin with a statement of Bayes Theorem, which is the central tool required for the application of Bayesian methods.

Let B_1, B_2,... be mutually exclusive and exhaustive events contained in a sample space S. (That is, $B_i B_j = \emptyset$, the empty set, for $i \neq j$, and $\cup_i B_i = S$.) Then for any event A,

$$\Pr(B_i|A) = \frac{\Pr(B_i)\Pr(A|B_i)}{\Pr(A)}$$

where

$$\Pr(A) = \sum_j \Pr(B_j)\Pr(A|B_j),$$

and $\Pr(A|B)$ denotes the conditional probability of A given B. This formula tells us how to transform the **unconditional** probabilities $\Pr(B_i)$ into the **conditional** probabilities $\Pr(B_i|A)$ which become relevant when it is known that the event A has occurred. The significance of this formula in statistics is that we can take A to be the event that the given sample values are observed, and let the B_i represent the unknown 'states of nature'.

We assume the set-up of Chapter 3, in that we have available a sample $x_1, \ldots, x_n$ of independent observations with common density function $f(x; \theta)$, where $\theta = (\theta_1, \ldots, \theta_m)$ is an unknown parameter vector. Suppose that the prior uncertainty regarding θ can be quantified in a probability distribution having density function $p(\theta)$, the **prior density** of θ. An application of Bayes Theorem then yields the **posterior density**

$$p(\theta|x) = \frac{p(\theta) \prod_{i=1}^{n} f(x_i; \theta)}{p(x)} \qquad (6.1)$$

(When x and θ are discrete this is immediate from Bayes formula; (6.1) is also valid when f and/or p are density functions of continuous random variables.) Since we may regard x as fixed, we can express Bayes formula simply as

$$p(\theta|x) \propto p(\theta)L(\theta)$$

where $L(\theta)$ is the likelihood function. Furthermore

$$p(x) = \int L(\theta)p(\theta)\,d\theta$$

is the (unconditional) joint probability density function of $x_1, \ldots, x_n$. The posterior density $p(\theta|x)$ quantifies the current beliefs about the parameter θ. Any probability statement about θ can now be made by an appropriate integration of $p(\theta|x)$. A suitable measure of central location of the posterior distribution will serve as a current estimate of θ.

Bayes formula is named after the Reverend Thomas Bayes, whose pioneering paper *An essay toward solving a problem in the doctrine of chances* was published (posthumously) in 1763. In that paper, Bayes proposed a method of making inferences about the parameter θ in a Binomial distribution, and effectively used formula (6.1) with $p(\theta) = 1$. Thus he assumed that ignorance about the success probability θ could be represented by a uniform prior distribution on the interval $[0, 1]$. The general form of the theorem was stated by Laplace in 1774.

We illustrate some important features of the Bayesian approach by means of a simple example.

Example 6.1

Consider again the aircraft component data given in Chapter 2, section 2.10 for Example 2.2. There are thirteen observations, three of which are Type II censored. For the purpose of illustration, we assume here that an exponential failure time distribution is appropriate; that is, the survivor function is given by $S(t) = e^{-\lambda t}$, where λ^{-1} is the mean time to failure. From equations (2.2) and (3.1) the likelihood function is given by

$$L(\lambda) = \lambda^r \exp\left\{-\lambda \sum_{i=1}^{r} t_i\right\} \exp\left\{-\lambda \sum_{i=r+1}^{n} t_i\right\} = \lambda^r e^{-s\lambda} \qquad (6.2)$$

where n is the number of observations, r the number of uncensored cases and $s = \sum_{i=1}^{n} t_i$. In the present example, we have $r = 10$ and $s = 23.05$.

The prior distribution
In order to determine the prior distribution of λ, it is necessary to quantify uncertainty about the failure rate, λ. This may be based on historical data on failures of similar components, or may be a subjective quantification based on relevant experience. Here it will be particularly convenient if we are able to use a prior distribution for λ with a density of the form

$$p(\lambda) \propto \lambda^{a-1} e^{-b\lambda} \tag{6.3}$$

which is the density of a Gamma(a, b) distribution. The reason for this is that the density (6.3) is of the same functional form as the likelihood (6.2), and will combine with $L(\lambda)$ in a simple way. The class of Gamma distributions here is called the **natural conjugate family** of distributions of λ. In order to determine the constants a and b, two independent pieces of information about the prior distribution of λ are required. For example, suppose λ is initially believed to be about 0.5, and unlikely to be less than 0.2. If we take the prior mode to be 0.5 and specify that $P(\lambda < 0.2) \approx 0.05$, then it can be seen that the values $a = 3$, $b = 4$ will be appropriate.

The posterior distribution
It now follows on multiplication of (6.2) and (6.3), and on substituting the numerical values, that

$$p(\lambda|t) \propto \lambda^{12} e^{-27.05\lambda}.$$

The constant of proportionality is found using $\int_0^\infty p(\lambda|t)\,d\lambda = 1$, and is therefore given by

$$\left\{ \int_0^\infty \lambda^{12} e^{-27.05\lambda}\,d\lambda \right\}^{-1} = 27.05^{13}/12!$$

Thus the posterior distribution of λ is a Gamma distribution, with parameters 13 and 27.05. Notice that this has the same functional form as the prior density, since the latter was chosen from the natural conjugate family. The class of Gamma distributions here is said to be **closed under sampling**. The prior and posterior densities of λ are illustrated in Figure 6.1. The posterior mode of λ is $12/27.05 = 0.444$, while the posterior mean is $13/27.05 = 0.481$; either of these values could be used as a point estimate of θ. (A more formal approach to estimation is outlined in section 6.5.) The posterior standard deviation is 0.133, which may be compared with the prior standard deviation of 0.175.

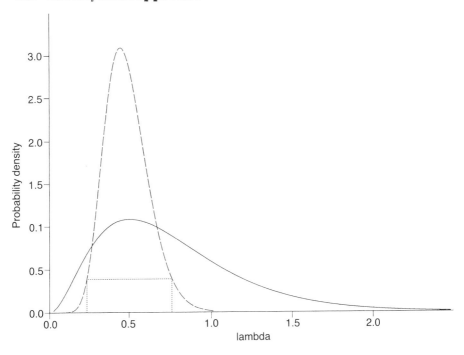

Figure 6.1 Prior (——) and posterior (– – –) densities of λ and 95% HPD interval ($\cdots\cdots$) for Example 6.1.

An alternative convenient way of summarizing the current uncertainty about λ is to supply one or more ranges of likely values of λ. The usual method is to give regions of **highest posterior density**. A highest posterior density (HPD) region is one within which the posterior density is always greater than at any value outside the region. In the present case, HPD regions are intervals. For example, the 95% HPD interval for λ is 0.231 to 0.758, and is shown in Figure 6.1. These values are either obtained numerically, or else read from tables for Bayesian statistics (for example, Isaacs *et al.* (1974)).

The 95% HPD region should be carefully distinguished from a 95% confidence region. The latter region is a random set of values within which the true value of λ will lie 95% of the time in repeated sampling. The 95% HPD region, on the other hand, has a direct probability interpretation: the unknown value of λ is believed to lie within the given region with probability 0.95. This assertion cannot be made for a confidence region, although such statements are often loosely made.

Probability statements about, and estimates of other parametric functions can be obtained in a similar way. We might be interested, for example, in the reliability $S(t_0)=e^{-\lambda t_0}$ at a specified time t_0. Since $S(t_0)$ is a function of λ, the posterior distribution of this quantity can be inferred from that of λ. In particular, suppose we wish to specify a likely maximum value for $S(t_0)$ on the basis of the sample data. From the posterior distribution of λ we find, for example, that $\Pr(\lambda < 0.2255|t) = 0.01$. Thus the posterior probability that $S(t_0)$ exceeds $e^{-0.2255t_0}$ is 0.01.

Predictive distributions

Statements about a future failure time S can be made by calculating the predictive distribution of S; that is, the distribution of S conditional on the observed failure times $t_1, \ldots, t_n$. This is obtained by an integration over λ,

$$p(s|t) = \int p(s|t, \lambda) p(\lambda|t) \, d\lambda$$

$$= \int p(s|\lambda) p(\lambda|t) \, d\lambda$$

since S is conditionally independent of T, given λ. Treating terms not involving s as constants, this gives

$$p(s|t) \propto \int_0^\infty (\lambda e^{-s\lambda})(\lambda^{12} e^{-27.05\lambda}) \, d\lambda$$

$$= \int_0^\infty \lambda^{13} e^{-(27.05 + s)\lambda} \, d\lambda$$

$$\propto (1 + 0.037s)^{-14}.$$

The constant of proportionality, obtained by integration, is 0.481. The **predictive survivor function** is given by

$$P(S > s|t) = \int_s^\infty p(s'|t) \, ds' = (1 + 0.037s)^{-13}.$$

Vague prior knowledge

If only weak prior knowledge about λ is available, it may be difficult to formulate the prior distribution of λ with any degree of precision. Moreover, two individuals might formulate their vague prior knowledge in quite different ways, and it is therefore important to investigate the behaviour of the posterior distribution under such circumstances. Suppose, for example, that the weak prior knowledge is represented by a Gamma(0.5, 0.5) distribution; here a 95% prior-probability range of values for λ is 0.001 to 5.024. Combining this prior distribution with the likelihood (6.1) yields a Gamma(10.5, 23.55) posterior distribution, which has mean 0.446 and standard deviation 0.138. These values may be compared with the previous values of 0.481 and 0.133 respectively. An important point here is that alternative prior specifications, provided they express fairly vague prior know-ledge in relation to the information conveyed by the data, will lead to similar posterior distributions. The reason is that such prior densities will often be approximately constant over the range of the parameter space where the likelihood function is appreciable. Thus it is very often unnecessary to formulate a prior distribution corresponding to vague prior knowledge with any great deal of precision.

In view of the above phenomenon, it is quite common to use a **non-informative reference prior**, which is often taken to be an improper limit of the appropriate family of conjugate distributions. Here, if we let a and b both tend to zero, every highest prior density interval tends to $(0, \infty)$, which is an intuitively appealing representation of 'no prior information'. This yields a Gamma(10,23.05) posterior distribution, which has mean 0.434 and standard deviation 0.137, again close to the values obtained previously. Formally, this is equivalent to working with a prior density of the form $p(\lambda) \propto \lambda^{-1}$, which is not a proper probability density, as its integral over $(0, \infty)$ is infinite. It is important to understand the interpretation of such a prior; we are not saying that this prior actually represents our prior belief, only that the posterior distribution obtained from it via a formal application of Bayes Theorem will hardly differ from our 'actual' posterior distribution when prior knowledge is very weak.

The above example illustrates some basic features of the Bayesian approach to inference. For details of the application of the Bayesian approach to many statistical models, the reader can consult the introductory texts by Lindley (1965) or Lee (1989). Martz and Waller (1982) discuss Bayesian analysis in reliability, including detailed discussion of the principal reliability models. This book also includes an extensive reference list to the literature on the application of Bayesian methods in this area.

6.4 FURTHER TOPICS IN BAYESIAN INFERENCE

In this section we discuss a number of general issues concerning Bayesian inference, with particular reference to reliability data analysis.

Prior specifications
In many cases it is not necessary to specify the prior distribution too precise-ly in a Bayesian analysis for the reasons given at the end of the previous section. In some cases, however, the posterior distribution may be quite sensitive to certain aspects of the prior distribution, which therefore require more accurate assessment. The elicitation of prior information requires

careful treatment, and any sensible approach should include checks on inconsistencies in prior specifications. Various techniques exist for accurate and consistent assessment of prior beliefs; see, for example, the text by Smith (1988).

As was demonstrated in Example 6.1, it is very convenient to use conjugate priors, and these should be used whenever possible. There is no reason why one's actual prior should be of this form, however there are circumstances when a conjugate prior is not appropriate. If, for example, we believed that the prior density in Example 6.1 were bimodal, then the form (6.3) would be clearly unsuitable. Very often, however, it is possible to represent the prior distribution by a mixture of conjugate priors, which also turns out to be mathematically convenient.

There is an extensive literature on appropriate forms for non-informative reference priors. For practical purposes, the differences in answers obtained under alternative choices are usually very small, for the reasons outlined above. A more important question is whether one should be using a non-informative prior at all in a particular application. Very often the non-informative prior approach gives a very satisfactory answer, but sometimes, especially in problems involving many parameters, blind application of such priors can lead to absurdities. In such cases there is no alternative to a careful consideration of prior specifications, and usually it will be desirable to report the results of the investigation under a range of reasonable priors.

In reliability analysis, a specific area where it is essential to work with realistic prior distributions is in the assessment of component reliability using component and/or systems data. In such problems, the prior specifications for component reliability may play a crucial rôle in a Bayesian analysis; see for example Mastran and Singpurwalla (1978) and David (1987). Further discussion of this topic appears in Chapter 9, section 9.3.

Finally, we mention the problem of forming a single prior distribution for a group of individuals with different prior beliefs. Lindley and Singpurwalla (1986a) discuss the pooling of expert opinion in a reliability analysis.

Nuisance parameters

Parameters involved in a model which are not considered to be of primary interest are termed **nuisance parameters**. In order to make inferences about the current parameter(s) of interest, it is necessary to eliminate the nuisance parameters. In principle, this is easily accomplished in a Bayesian analysis, since if $p(\psi, \phi|x)$ is the joint posterior density of the parameter of interest, ψ, and the nuisance parameter, ϕ, then probability statements about ψ alone are obtained from the **marginal** posterior density $p(\psi|x)$ of ψ given by

$$p(\psi|x) = \int p(\psi, \phi|x) \, d\phi.$$

Thus nuisance parameters are eliminated simply by 'integrating them out' of the joint posterior density.

Returning to Example 6.1, suppose that instead of an exponential failure time distribution, a two-parameter Weibull distribution with survivor function $S(t) = e^{-\lambda t^{\gamma}}$ is assumed. The likelihood (6.2) becomes

$$L(\lambda, \gamma) = (\lambda\gamma)^r \prod_{i=1}^r t_i^{\gamma} e^{-\lambda s(\gamma)}$$

where

$$s(\gamma) = \sum_{i=1}^n t_i^{\gamma}.$$

Now assume a prior for λ of the form (6.3), an arbitrary prior $p(\gamma)$ for γ and, for simplicity, assume prior independence of λ and γ. Then the joint posterior density of λ and γ is

$$p(\lambda, \gamma | t) = p(\gamma) \lambda^{a+r-1} \gamma^r \prod_{i=1}^r t_i^{\gamma} e^{-\lambda s(\gamma)}. \tag{6.4}$$

The marginal posterior density of γ is now obtained via integration of (6.4) with respect to λ. Using the substitution $u = \lambda s(\gamma)$, this is found to be

$$p(\gamma | t) \propto p(\gamma) \left\{ \gamma^r \prod_{i=1}^r t_i^{\gamma} / s(\gamma)^{a+r} \right\}.$$

The expression in curly brackets is referred to as the **integrated likelihood** of γ.

Finally, we mention again the danger of assigning vague prior distributions for the nuisance parameter(s) ϕ. It might be the case that the posterior distribution of the parameter(s) of interest, ψ, depends rather critically on the prior adopted for ϕ, especially when the data contain little information on ϕ.

Bayes factors
The hypothesis testing problem can be approached from a Bayesian viewpoint. A hypothesis H about θ specifies a region ω of the parameter space in which θ is supposed to lie. In Bayesian statistics, one can calculate directly the posterior probability that the given hypothesis H is true. One could then reject H if $\Pr(H|x) \leq \alpha$, where α is some specified small number. In Example 6.1, suppose we wished to test the hypothesis $H: \lambda < 0.25$. Then the posterior probability of H can be obtained by integrating the posterior density (with its constant of proportionality) over the range $[0, 0.25]$, or by referring to tables

of the Gamma distribution function. It is found that $\Pr(H|t)=0.02$, so H is unlikely to be true.

Equivalently, it is often useful to consider the magnitude of the **posterior odds ratio**, given by

$$\frac{\Pr(H|x)}{\Pr(\bar{H}|x)} = \frac{\Pr(H)}{\Pr(\bar{H})} \frac{p(x|H)}{p(x|\bar{H})} \tag{6.5}$$

where $\bar{H}$ means 'not H'. In words,

posterior odds = prior odds × 'Bayes factor'.

The **Bayes factor** is the relative probability of observing the data under H and under $\bar{H}$. A value greater than one corresponds to support for H from the data. Alternatively, the consequences of making incorrect decisions can be formally taken into account by using a decision-theoretic approach; this is reviewed in the next section. A particular application is to **reliability demonstration testing**, which is concerned with the question of how much testing should be carried out in order to demonstrate a required level of reliability with a specified probability; see Martz and Waller (1982), for example. The application to assessment of competing models is discussed in section 6.6.

The relation (6.5) is particularly useful when testing a point hypothesis of the form $H: \theta = \theta_0$. In such cases it is necessary to assign a positive prior probability to the hypothesis, for otherwise the posterior probability that $\theta = \theta_0$ will always be zero! In practice, if the sharp hypothesis $\theta = \theta_0$ is not really plausible, it may be more realistic to consider an hypothesis of the form $|\theta - \theta_0| < \delta$, where δ is some small tolerance value. On the other hand, if there is appreciable prior probability in the region $|\theta - \theta_0| < \delta$, it may be convenient to use as an approximation a prior which assigns a positive weight to the value θ_0. When testing a point hypothesis, it is not possible to use a non-informative reference prior for θ under $\bar{H}$, for otherwise the data will always lead to acceptance of H (since the extreme uncertainty about θ under $\bar{H}$ makes **any** observed data extremely unlikely).

Returning to Example 6.1, consider testing the hypothesis $H: \lambda = 0.25$. The Bayes factor here turns out to be 0.81; see, for example, Berger (1985) for the computation of Bayes factors. Thus our belief in the value of 0.25 for λ would be slightly reduced in the light of the data. Although it would not have been possible to use the non-informative prior $p(\lambda) \propto \lambda^{-1}$ here, there are some devices which appear to give reasonable answers when prior knowledge is weak; see, for example, Spiegelhalter and Smith (1982).

Bayesian computation

It is clear from Example 6.1 that the derivation of posterior quantities of interest, such as HPD regions, posterior moments, marginal posterior distributions and predictive distributions, requires integration techniques. In certain problems the integration may be carried out analytically, and closed-form expressions obtained. Very often, however, it is not possible to obtain closed-form expressions for the desired integrals, and either numerical integration or analytic approximation is required. Returning to the discussion of nuisance parameters for Example 6.1, if λ were considered to be the parameter of interest, then analytic integration of (6.4) over the Weibull shape parameter γ is not possible. The computational problems are particularly severe in examples with many nuisance parameters, as these require high-dimensional numerical integration. This important general area is the subject of much current research. For a description of some existing Bayesian computer software which includes numerical integration routines, see Goel (1988). (This article also gives details of programs specifically aimed at reliability analysis.)

The simplest approximations to posterior quantities arise by appealing to large-sample theory. In Chapter 3 it was stated that, in large samples, the ML estimator $\hat{\theta}$ of a parameter θ is approximately normally distributed with mean θ and variance-covariance matrix $V = J^{-1}$, where J is the observed information matrix defined in that chapter. The corresponding Bayesian result is that, under suitable regularity conditions, the posterior distribution of θ is approximately normal, with mean $\hat{\theta}$ and variance-covariance matrix V, provided the prior density of θ is continuous and positive throughout the parameter space. The reason for this is that in large samples the posterior distribution becomes dominated by the likelihood function, and the log-likelihood becomes approximately quadratic. Thus, at least in regular problems with large samples, the classical and Bayesian approaches lead to essentially the same inferences, although interpretation is of course quite different.

Consider again Example 6.1. From Chapter 3, section 3.3 the ML estimate of λ is $\hat{\lambda} = r/s = 0.434$, and the observed information $J = s^2/r = 53.130$. The approximate posterior mean and standard deviation of λ are therefore $\hat{\lambda} = 0.434$ and $J^{-1/2} = 0.137$. Note that, in this application, these values actually coincide with the exact values obtained using the non-informative prior $p(\lambda) \propto \lambda^{-1}$. The approximate 95% HPD limits obtained from the Normal approximation to the posterior distribution of λ are $0.434 \pm 1.96 \times 0.137$, giving the interval 0.17 to 0.70. This is, of course, identical with the approximate confidence interval for λ obtained by method (2.) of section 3.4 of Chapter 3. Unless the sample size is very large, however, HPD intervals based on Normal approximations to posterior distributions can be rather sensitive to the choice of parametrization. It is generally preferable to base the approximation on the likelihood-ratio criterion, given as method (1.) in Chapter 3, section 3.4. This method yields the interval 0.22 to 0.76, which is almost identical with the exact 95% HPD interval given in section 6.3.

As is the case in frequentist statistics, standard ML approximations may be too crude in certain cases (such as for heavily censored data often arising in reliability studies). More accurate approximations have been given by Tierney and Kadane (1986), which we reproduce here, as they provide useful and easily implemented tools for posterior analysis. Further details and examples of the use of these formulae in Bayesian computation appear in Kass *et al.* (1988).

Write $\theta = (\psi, \phi)$ and consider approximation of the marginal posterior density of ψ. Here both ψ and ϕ may be vector-valued. Write $L_1(\theta) = p(\theta)L(\theta)$, and let $\hat{\phi}_\psi$ be the conditional posterior mode of ϕ given ψ; that is, the value of ϕ which maximizes $L_1(\theta)$ for fixed ψ. Let $\Sigma(\theta)$ be the negative inverse of the Hessian (matrix of second-order derivatives with respect to ϕ) of $\log L_1(\psi, \cdot)$. Then, approximately,

$$p(\psi|x) \propto L_1(\psi, \hat{\phi}_\psi)|\Sigma(\psi, \hat{\phi}_\psi)|^{1/2}$$

where $|A|$ denotes the determinant of A. Alternatively, and to the same order of approximation, in the above formula $\hat{\phi}_\psi$ may be taken to be the ML estimate of ϕ for given ψ, and $\Sigma(\theta)$ the inverse of the submatrix of the observed information J corresponding to ϕ. Interestingly, this latter form has connections with various modified forms of profile likelihood used for the elimination of nuisance parameters in the classical approach; see Sweeting (1987b), for example.

Consider next the approximation of posterior moments. Define $L_1(\theta)$ as above and let $L_1^*(\theta) = g(\theta)L_1(\theta)$. Let $\hat{\theta}, \theta^*$ be the parameter values which maximize $L_1(\theta)$ and $L_1^*(\theta)$ respectively, and $\Sigma(\theta), \Sigma^*(\theta)$ be the negative inverse of the Hessian of $\log L_1(\theta)$, $\log L_1^*(\theta)$ respectively. Then if $g(\theta)$ is a positive function, one has the following approximation to the posterior expectation of $g(\theta)$.

$$E\{g(\theta)|x\} = \frac{L_1^*(\theta^*)|\Sigma^*(\theta^*)|^{1/2}}{L_1(\hat{\theta})|\Sigma(\hat{\theta})|^{1/2}}. \qquad (6.6)$$

Finally we note that ML-type approximations may be inapplicable for some models. For example, a number of models in reliability include a guarantee, or threshold parameter (see Chapter 2, section 2.7). For such models the regularity assumptions required for these approximations do not hold (*cf.* the discussion in section 3.5 of Chapter 3). One important model is the three-parameter Weibull distribution. The practical advantage of adopting a Bayesian approach when interest centres on the threshold parameter has been argued by Smith and Naylor (1987). In this case the posterior distribution of the threshold parameter is usually markedly non-Normal.

6.5 DECISION ANALYSIS

In many reliability problems, the purpose of collecting the reliability data is to help in solving a decision problem. In practice, the decision-making process is sometimes carried out in an informal manner on the basis of the results of the statistical analysis. A more formal approach, however, requires an assessment of the costs and consequences of making alternative decisions under the different 'states of nature'. These states of nature may be alternative values of a critical parameter θ, or perhaps alternative reliability models. In a decision analysis the various costs and consequences are quantified as **losses**.

In general, one specifies a **decision space** D of possible decisions and a **loss function** $L(d, \theta)$ which measures the cost if decision d in D is taken when θ is the true state of nature. In the Bayesian approach to decision analysis, one chooses that decision d^* which minimizes $E[L(d, \theta)|x]$, the **posterior expected loss**, or **Bayes' risk**, under decision d. The optimal decision d^* is called the **Bayes' decision**. Thus the Bayesian analysis of failure-time data just becomes an ingredient of a full decision analysis. Note that losses need not be assessed solely in terms of monetary units; in general, losses should be in units of negative personal **utility**. A recent applied text on Bayesian decision analysis is Smith (1988).

As an example, a common problem in reliability is the derivation of a safe inspection interval, when inspection is to be introduced in order to detect small cracks in a structure or component, for example. Letting τ denote the length of time between inspections, the decision space D might consist of all positive values of τ, or six-monthly intervals, for example, in order to correspond to an established regular maintenance cycle. The overall loss if a component fails between inspections will take into account the cost in monetary terms of damage to the system, as well as the possible environmental and human costs involved. On the other hand, there will of course be costs associated with carrying out inspections. In the area of structural reliability, a Bayesian approach to optimal reliability-based design is natural, since it may be necessary to incorporate subjective judgement into the assessment of load and resistance variables; see Cornell (1972) for details.

An example: optimal replacement

Returning to Example 6.1, suppose that, on the basis of these data, it is required to determine a suitable planned replacement time for this component. Let N denote the planned component replacement time, and consider the replacement strategy which makes a planned replacement when the age of the component reaches N, and a service replacement whenever a failure occurs before N. Let T be the failure time of a component. Assume that when $T > N$ a cost c_1 associated with planned replacement will be incurred, and when $T \leq N$ there will be a service replacement cost $c_2 > c_1$.

We first review the situation when the failure-time distribution is known. Full details may be found in Cox (1962a), Chapter 11. Consider a very large number m of components. The total expected cost of replacement is

$$mS(N)c_1 + mF(N)c_2$$

where $S(t)$ and $F(t)$ are the survivor and distribution functions respectively of T. The expected length of time for which the m components are in use is

$$m\left\{\int_0^N tf(t)\,dt + NS(N)\right\} = m\int_0^N S(t)\,dt$$

on integration by parts. The mean cost per unit time is therefore

$$C = (c_2 - \{c_2 - c_1\}S(N))\bigg/\int_0^N S(t)\,dt \qquad (6.7)$$

The optimal replacement time, N^*, is now found by minimizing this expression. Here we are assuming that utility is adequately measured by negative monetary loss; this will usually be the case when the financial consequences of alternative decisions are not particularly severe. We note that if the failure-time distribution is exponential then $N^* = \infty$, which implies that the optimal policy is to make service replacements only. If the failure-time distribution is Weibull with shape parameter $\eta > 1$, however, a finite number N^* exists which minimizes (6.7).

The above analysis assumes that the failure-time distribution F is known, and the question which now arises is how to proceed on the basis of incomplete knowledge of F. Denoting the unknown parameters of F by θ, the loss, $L(N, \theta)$, when the planned replacement time is N and θ is the true parameter value, is given by equation (6.7) under the failure-time distribution indexed by θ. Given the failure-time data of Example 2.2, in section 2.10 of Chapter 2, the Bayes replacement policy is obtained by combining the prior and sample information to obtain the posterior distribution of θ, and then minimizing the posterior expected loss $E\{L(N, \theta)|t\}$. The exact solution to this problem will normally require numerical integration and minimization techniques. For sufficiently large samples, however, $L(N, \hat{\theta})$ may be used as an approximation to $E\{L(N, \theta)|t\}$. Alternatively, the more accurate approximation given by (6.6) may be used.

Application to statistics
The general statistical problems of estimation and hypothesis testing can be tackled using a Bayesian decision-theoretic approach. In the estimation problem, an estimator $T(x)$ of θ is regarded as a decision function; the optimal estimate

$T^*(x)$ of θ, given data x, is that value t which minimizes the posterior expected loss $E[L(t, \theta)|x]$. The most common choice of loss function here is a quadratic loss function of the form $L(t, \theta) = c(t - \theta)^2$. It can be shown that under quadratic loss the optimal Bayes estimate is the posterior mean. Other choices of loss function may lead to different estimators; for example, the absolute error loss $L(t, \theta) = c|t - \theta|$ leads to the posterior median as the optimal estimate.

In the hypothesis testing problem, a formal decision-theoretic solution requires the formulation of the various losses associated with making incorrect decisions. Let d_1, d_2 be the decisions corresponding to acceptance and rejection respectively of a hypothesis H. Consider the following loss structure. When H is true, $L(d_1, \theta) = 0$, $L(d_2, \theta) = K_1$, and when H is false $L(d_1, \theta) = K_2$, $L(d_2, \theta) = 0$ ('0–K_i' loss). Then the posterior expected losses under d_1, d_2 are simply $K_2 \Pr(\bar{H}|x)$, $K_1 \Pr(H|x)$ respectively. Thus decision d_2 should be chosen (that is, 'reject H') when $K_1 \Pr(H|x) \le K_2 \Pr(\bar{H}|x)$ or, in terms of the Bayes factor introduced in section 6.4,

$$\frac{p(x|\bar{H})}{p(x|H)} \ge \frac{K_1 \Pr(H)}{K_2 \Pr(\bar{H})}.$$

Finally, we point out that there are very strong connections between Bayesian and classical decision theory, as expounded by Wald (1950). For details of this, and the general theory surveyed in this section, see for example De Groot (1970) or Berger (1985).

6.6 BAYESIAN ANALYSIS OF RELIABILITY DATA

In this section we consider some aspects of the Bayesian analysis of accelerated life models, defined in Chapter 4, and other parametric and semiparametric reliability regression models. As mentioned in section 6.4, in regular cases, large-sample ML analyses of general parametric models are formally equivalent to Bayesian analyses with relatively vague prior information. Thus, to a certain extent, the ML analyses performed in Chapters 3 and 4 can be given approximate Bayesian interpretations. In practical applications, the appropriateness of ML methods as approximations can be assessed by investigating the extent to which the log-likelihood surface is quadratic.

For example, for the failure stress data given in Table 4.2, the maximum likelihood estimate 0.58 of α obtained in section 4.8 under model M_1 is an approximation to the posterior mean of α, provided no substantial prior information concerning α is available. Similarly, the estimated standard error of $\hat{\alpha}$ obtained via the maximum likelihood procedure is an approximation to the posterior standard deviation of α, and approximate 95% confidence intervals for α are also approximate 95% HPD intervals for α. When substantial prior information on one or more of the parameters is available in the form of a Normal distribution, it is possible to combine this with the large-sample approximations in a simple way, since both likelihood and prior will be of Normal form.

The significance levels obtained in the various hypothesis tests applied in section 4.8 do not however correspond to direct Bayesian calculation of posterior odds ratios, the latter depending crucially on prior assumptions. It is usually possible, however, to give a Bayesian interpretation in the following way: reject the hypothesis H at level α (0.05, for example) if the parameter value specified by H falls outside the Bayesian $100(1 - \alpha)\%$ HPD region for that parameter. With this interpretation, acceptance of H does not imply that H is believed to be true, but merely that the data do not appear to contradict the hypothesis H, as is the case in the classical framework.

A somewhat different type of analysis may be applied for the Kevlar 49 failure data given in Table 4.4. A possibly more realistic approach to the specification of the joint prior distribution of $\psi_1, \ldots, \psi_8$ would be to regard the eight spools as essentially indistinguishable, prior to observing the data: they could be regarded as a sample from a larger population of spools, for example. The prior distribution of $\psi_1, \ldots, \psi_8$ is then taken to be exchangeable, meaning that the form of the prior adopted is independent of the ordering of the spools. A prototype model of this kind is the **Normal exchangeable model** where, conditionally on the parameter values, the responses are Normally distributed and, conditionally on certain hyperparameters, the parameters themselves are assumed to be jointly symmetrically Normally distributed. A suitable prior distribution is then assigned to the hyperparameters; see for example Lindley and Smith (1972) for details. In the present case the basic model is non-Normal, but one could still use Lindley and Smith's approach to specification of the prior. One feature of this approach is that the estimates of the ψ_i become pulled towards their average value. Similar procedures can be justified from a classical point of view, and the estimators are referred to as **shrinkage** or **Stein-type** estimators.

Maximized likelihoods were used in sections 4.8 and 4.9 for choosing between alternative competing models. These can also be used as a rough guide in Bayesian model assessment. Generally, if M_1, M_2 are two competing reliability models, the posteriors odds ratio is given by

$$\frac{\Pr(M_1|x)}{\Pr(M_2|x)} = \frac{\Pr(M_1)}{\Pr(M_2)} \frac{p(x|M_1)}{p(x|M_2)}.$$

When M_1 and M_2 are of location-scale form, the sample size is large and prior knowledge vague then, as a first approximation, the final ratio may be replaced by the maximized likelihood ratio.

When the sample information is not great, large-sample approximations may be poor, as mentioned in the previous section. In the analysis of reliability data, this may occur when the sample size is small, or when the data contain many censored observations, or when there are a large number of regressor variables, or a combination of these factors. As discussed in section 6.4, exact computations will usually involve high-dimensional numerical integration. On the other hand, the more accurate approximations of section 6.4 may be applicable.

In the case of location-scale regression models, multivariate Student t and χ^2 distributions often provide quite accurate approximations to posterior distributions; see Sweeting (1984, 1987a and 1988). In the case of uncensored data the Bayesian analysis of a location-scale model using a vague prior has a (conditional) frequentist interpretation, so that computational or approximate methods in this case have a wide appeal; see Lawless (1982) for further details.

We illustrate these ideas using the censored failure data given by Crawford (1970), which have been analyzed by a number of authors. In particular, Naylor and Smith (1982) carry out a Bayesian analysis of these data using exact numerical integration by quadrature methods. The data, given here in Table 6.1, arise from temperature accelerated life tests on electrical insulation motorettes. Ten motorettes were tested at each of four temperature levels, resulting in 17 failed units and 23 unfailed (i.e. censored) units.

Table 6.1 Life test data on motorettes

Insulation life in hours at various test temperatures (* denotes a censored time)			
150°C	170°C	190°C	220°C
8064*	1764	408	408
8064*	2772	408	408
8064*	3444	1344	504
8064*	3542	1344	504
8064*	3780	1440	504
8064*	4860	1680*	528*
8064*	5196	1680*	528*
8064*	5448*	1680*	528*
8064*	5448*	1680*	528*
8064*	5448*	1680*	528*

We fit a model of the form

$$\log(\text{failure time}) = \alpha + \beta \, (\text{temperature} + 273.2)^{-1} + \sigma z$$

where time is in hours and temperature in °C. Such models have been found satisfactory for many products undergoing temperature accelerated life testing; see, for example, Nelson and Hahn (1972). Figures 6.2 and 6.3 show the exact

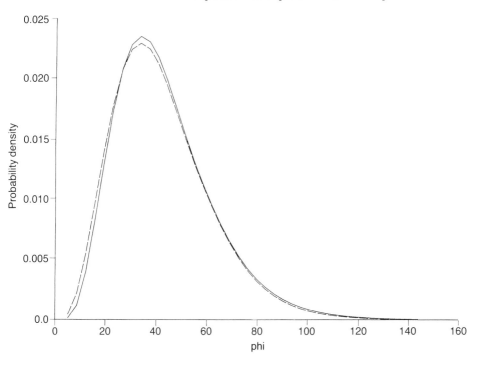

Figure 6.2 Exact (——) and approximate (– – –) posterior densities of ϕ for motorettes life test data.

and approximate posterior distributions of the precision parameter $\phi = \sigma^{-2}$, and the regression parameter β, assuming a standard Gumbel distribution for the error term z. The approximations used here are those described in Sweeting (1987a), and are simply χ^2 and Student t distributions respectively, each having approximately 10 degrees of freedom. The exact densities were obtained by Gaussian quadrature methods. Considering the degree of censoring, these approximations are remarkably accurate here. (If a standard Normal distribution is assumed for z, as is done in Nelson and Hahn (1972), the posterior approximations turn out slightly better.)

We conclude this section by describing Bayesian approaches to the analysis of the semiparametric regression models discussed in Chapter 5. One simple approach is via the **marginal likelihood** interpretation of (5.3) described in that chapter. If the likelihood (5.3) is combined with the prior density of β in the usual way, then the result will be the posterior density of β, conditional on the ranks of the observations, ignoring the actual lifetimes.

A more difficult problem is that of finding the full posterior distribution of β given the observed lifetimes. Here the entire baseline hazard function h_0 must be regarded as an unknown parameter. In order to obtain the posterior

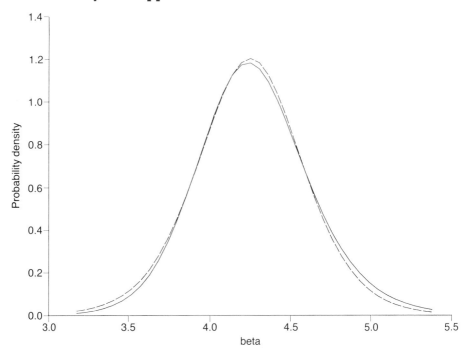

Figure 6.3 Exact (——) and approximate (– – –) posterior densities of β for motorettes life test data.

distribution of β, one needs to integrate out the nuisance parameter h_0 from the joint posterior distribution of β and h_0. One is therefore led to consider prior distributions for unknown functions, which clearly involves a higher level of mathematical analysis, not to mention the difficulties involved in eliciting prior beliefs about unknown functions. It is of some interest however to investigate the consequences of assuming alternative forms of prior for h_0.

When prior knowledge about h_0 is vague, one would like to be able to use a suitable reference prior distribution for h_0. On the other hand, in cases where one has quite strong prior beliefs about the form of the hazard function, such an approach would allow a degree of uncertainty to be introduced into a full parametric specification for h_0. In the case of vague prior knowledge about h_0, it turns out that alternative reasonable forms of prior lead to marginal posterior distributions for β which, to a first approximation, correspond to using (5.3) as likelihood for β. In addition, the Kaplan–Meier estimate of the survivor function $S(t)$ emerges as an approximation to the posterior mean of $S(t)$, under vauge prior knowledge about h_0. We refer the reader to Kalbfleisch and Prentice (1980) for details.

7

Multivariate models

In some cases there is more than a single failure time to be recorded for a system. There might be times to failure of different types, or of different components, or of partial failures or other associated events. In this chapter some models for, and some analyses of, data consisting of such multiple failure times are considered.

7.1 PRELIMINARIES

Suppose that there are p potential failure times $T_1, \ldots, T_p$ associated with a system. Some examples are as follows.

1. *Components in parallel.* The system continues to operate until all p components have failed. It may be possible to observe some or all of the individual component failure times before final failure, which occurs at time $V = \max(T_1, \ldots, T_p)$.
2. *Components in series.* The system fails as soon as any one component fails, this being at time $U = \min(T_1, \ldots, T_p)$. Often, only U and the failing component number C are observable, so that $U = T_C$, the other Ts being then right-censored at U.
3. *Competing and complementary risks.* System failures may be associated with one of p different causes, types or modes, so T_j represents the time to system failure from cause j. The observations comprise (C_S, U_S), the cause and time of the sth failure, for $s = 1, \ldots, r$. The case $r = 1$ has been studied under the title 'Competing Risks' as in section 7.4 and the case $r = p$ under 'Complementary Risks'. There are many variations, with repairable and non-repairable systems, and varying degrees of completeness in the observations, e.g. sometimes the causes of failure are unobserved.

It will be seen even from this rather restricted list that the variety of potential observation patterns is wide. The main developments in statistical methodolgy have been for two cases

1. where all of the C_j and U_j are observed, and
2. where only (C_1, U_1) is observed.

Sections 7.3 and 7.4 concentrate on these. First, however, some basic notes on multivariate failure time distributions are given in section 7.2.

Much work in reliability theory involving complex systems is based on an assumption of independence, e.g. independent functioning of components. However, this is often not likely to be anywhere near true, particularly when there is load-sharing between components or randomly fluctuating loads, or generally where failure of one type affects the probability of failure of another type. The distributions noted below all contain the possibility of dependence between the T_j.

7.2 SOME MULTIVARIATE FAILURE TIME DISTRIBUTIONS

Let $T = (T_1, \ldots, T_p)$ be the vector of failure times with joint survivor function $S(t) = \Pr(T \geq t)$, where $t = (t_1, \ldots, t_p)$ and $T \geq t$ means $T_j \geq t_j$ for $j = 1, \ldots, p$. The corresponding joint density, assuming the T_j to be jointly continuous, is $(-1)^p \delta^p S(t) / \delta t_1 \ldots \delta t_p$. Independence of the T_js is defined by $S(t) = \Pi_{j=1}^p S_j(t_j)$ where $S_j(t_j) = \Pr(T_j \geq t_j)$ is the marginal survivor function of T_j. When the T_j are not independent the degree of dependence can be measured in various ways. One of these is based on the ratio $R(t) = S(t) / \Pi_{j=1}^p S_j(t_j)$; if $R(t) > 1$ for all t there is positive dependence, and vice versa. Lehmann (1966) discussed a similar ratio of distribution functions under the title 'quadrat dependence'. Shaked (1982) presents some more general ideas.

Various hazard functions and failure rates have found application in the literature. The conditional hazard function for T_j given $T \geq t$ (see section 3.2) is

$$h_j(t_j | T \geq t) = \lim_{\delta \to 0} \delta^{-1} \Pr(T_j \leq t_j + \delta | T \geq t) = -\partial \log S(t) / \partial t_j; \qquad (7.1)$$

Clayton (1978) uses this in the bivariate case. Johnson and Kotz (1975) propose a 'vector multivariate hazard rate' $h(t) = -\nabla \log S(t)$ of which (7.1) is the jth component; ∇ here denotes the vector 'grad' operator. Brindley and Thompson (1972) say that the T-distribution is IFR/DFR (i.e. has multivariate increasing/decreasing failure rate, see section 3.2) if $Q_\Delta(t) = \Pr(T \geq t + \Delta 1_p | T \geq t)$ is decreasing/increasing in t for all $\Delta > 0$; 1_p here denotes $(1, \ldots, 1)$.

Indicators such as $R(t)$, $Q_\Delta(t)$ and correlations $\rho(T_j, T_k)$ carry useful information, but may be difficult to evaluate for specific distributions.

Bivariate exponential (Gumbel, 1960)
Gumbel suggested three different forms with joint survivor functions as follows.

1. $S(t) = \exp(-t_1 - t_2 - \delta t_1 t_2)$. The dependence parameter δ satisfies $0 \leq \delta \leq 1$, $\delta = 0$ yielding independence of T_1 and T_2. The ratio $R(t) = \exp(-\delta t_1 t_2)$ and the correlation $\rho(T_1, T_2)$ falls from 0 to -0.4 as δ rises from 0 to 1; thus there

is only negative dependence. The vector hazard rate is $h(t) = (1 + \delta t_2, 1 + \delta t_1)$, and the distribution is IFR since $\log Q_A(t) = -\Delta\{2 + \delta(t_1 + t_2 + \Delta)\}$.

2. $S(t) = e^{-t_1 - t_2}\{1 + \delta(1 - e^{-t_1})(1 - e^{-t_2})\}$. Here, $-1 \leq \delta \leq 1$ with $\delta = 0$ yielding independence. The ratio $R(t) = 1 + \delta(1 - e^{-t_1})(1 - e^{-t_2})$ and $\rho(T_1, T_2) = \delta/4$; there is positive (negative) dependence for $\delta > 0$ ($\delta < 0$).

3. $S(t) = \exp\{-(t_1^\delta + t_2^\delta)^{1/\delta}\}$. Here, $\delta \geq 1$ with $\delta = 1$ yielding independence. The ratio $R(t) > 1$ for $\delta > 1$, but $\rho(T_1, T_2)$ is difficult to evaluate.

These survivor functions have been given in standard form. General scaling is achieved by replacing t_j by $\lambda_j t_j$ or t_j/α_j. Corresponding multivariate Weibull forms can then be constructed by replacing t_j by $t_j^{\eta_j}$. The effect of covariates x can be accommodated via loglinear models $\log \lambda_j = x^T \beta_j$ ($j = 1, 2$) as in Chapter 4.

Bivariate exponential (Freund, 1961)
In a two-compartment system the lifetimes, T_1 and T_2, of the components are exponential but the failure of one component alters the stress on the other. The effect is to change the rate parameter of the surviving component from λ_j to λ_j' ($j = 1, 2$). The joint survivor function is

$$S(t) = (\lambda_1 + \lambda_2 - \lambda_2')\{\lambda_1 e^{-(\lambda_1 + \lambda_2 - \lambda_2')t_1 - \lambda_2' t_2} + (\lambda_2 - \lambda_2')e^{-(\lambda_1 + \lambda_2)t_2}\}$$

for $t_1 \leq t_2$; the form $t_1 \geq t_2$ is obtained by interchanging λ_1 and λ_2, λ_1' and λ_2', t_1 and t_2. Notice that the marginal distributions are not exponential so the joint distribution is not strictly a bivariate exponential. Independence of T_1 and T_2 obtains if and only if $\lambda_1 = \lambda_1'$ and $\lambda_2 = \lambda_2'$. The correlation $\rho(T_1, T_2)$ satisfies $-1/3 < \rho < 1$.

Bivariate exponential (Marshall and Olkin, 1966)
These authors derived the joint survivor function

$$S(t) = \exp\{-\lambda_1 t_1 - \lambda_2 t_2 - \lambda_{12} \max(t_1, t_2)\}$$

to describe a two-component system subject to randomly occurring shocks (in fact, governed by Poisson processes). The distribution is also the only one to have the property that, given survival of both components to age t, the residual lifetimes are independent of t, i.e.

$$\Pr(T_1 \geq t_1 + t, T_2 \geq t_2 + t | T_1 \geq t, T_2 \geq t) = \Pr(T_1 \geq t_1, T_2 \geq t_2);$$

this characteristic no-ageing property is well known for the univariate exponential. Another characterization of the distribution is that $T_1 = \min(X_1, X_2)$ and $T_2 = \min(X_2, X_3)$ where X_1, X_2 and X_3 are independent exponential variates;

it follows that $\min(T_1, T_2)$ is also exponential. The dependence parameter λ_{12} is restricted to non-negative values and the correlation $\rho(T_1, T_2) = \lambda_{12}/(\lambda_1 + \lambda_2 + \lambda_{12})$; thus $0 \leq \rho \leq 1$. A complication is that the distribution has a singular component, with non-zero probability concentrated on the line $t_1 = t_2$; in practice, data do not usually exhibit such singularities. Also, theoretical aspects of estimation are non-standard (Proschan and Sullo, 1976).

Weibull mixtures
These models arise when p-component systems have independent Weibull component lifetimes, but there is a certain random variation between systems. Such variation might be due to randomly differing conditions in their assembly, installation and operation. The T_j are initially taken to be independent with survivor functions $S_j(t_j) = \exp(-\lambda_j t_j^{\eta_j})$; thus T_j has log-mean $E[\log T_j] = -\lambda \eta_j^{-1}(\log \lambda_j + \gamma)$, where γ is Euler's constant. By analogy with normal linear mixed-effects models (in which the observation is expressed as a sum of deterministic factors and Normally distributed quantities arising from different sources of variation) Crowder (1985) suggested taking $\log \lambda_j = \log \xi_j + \log \lambda$, where ξ_j is a fixed-effects parameter and λ varies randomly over systems. The joint survivor function is then

$$S(t) = \int_0^\infty \exp\left(- \sum_{j=1}^p \lambda_j t_j^{\eta_j}\right) dG(\lambda)$$

where G is the distribution function of λ.

In Crowder (1985) a Gamma distribution with shape parameter v is assumed for λ, resulting in $S(t) = (1 + s)^{-v}$ where $s = \sum \xi_j t_j^{\eta_j}$. This is the survivor function of the multivariate Burr distribution with association parameter v. See also Takahasi (1965), Clayton (1978), Hougaard (1984) and Lindley and Singpurwalla (1986b). In Crowder (1989) a stable law of characteristic exponent v for λ is suggested, on physical grounds, with result $S(t) = \exp(-s^v)$. Watson and Smith (1985) and Hougaard (1986a, b) had previously derived this distribution. An extended form $S(t) = \exp\{\kappa^v - (\kappa + s)^v\}$ is also proposed, containing a further parameter κ with particular interpretation. These distributions have easily accessible hazard functions, marginal and conditional survivor functions, and other properties. Details, together with applications to real data, can be found in the papers cited.

Other multivariate distributions
A form commonly employed for statistical analysis is the multivariate lognormal, where the log T_j are jointly Normally distributed. However, for failure time data this has the disadvantage that the joint survivor and hazard functions are not expressible in closed form, thus making censored data likelihoods difficult to compute. Other multivariate distributions are reviewed in Johnson and Kotz (1972, Chapter 41).

7.3 COMPLETE OBSERVATION OF T

When all components of T are observed and identified, though some might be censored or missing in particular cases, statistical analysis follows the univariate methodology, either parametric or nonparametric. In the parametric case a multivariate distribution of specified form, such as one of those in section 7.2, is used as a basis for inference; this can involve informal and formal methods like those described in earlier chapters. For example, suppose that one of the bivariate exponential distributions were to be applied, with loglinearly modelled scale parameters λ_1 and λ_2. Then, informal methods include the plotting of the component failure times, t_1 and t_2, versus the xs as described in Chapter 4, and more formal analysis would be based on the likelihood function. Some examples will be given in this section.

Nonparametric and semiparametric methods are as yet less widely applied and this area will not be dealt with in any detail here; see Clayton (1978) and Oakes (1982).

Example 7.1

Table 7.1 gives the breaking strengths of parachute rigging lines. Eight lines on each of six parachutes were tested, and on each line the strength was measured at six equally spaced positions at increasing distances from the hem. The line number indicates its serial place round the parachute. The parachutes have been exposed to weathering for different periods of time, though these are unrecorded, and this explains the obvious overall differences in strength between them. The primary concern is to determine how breaking strength varies with line and position.

The data here are very tidy, comprising a complete factorial lay-out with no missing or censored values. It is thus tempting to call up a standard statistical package, and so the MANOVA program in SPSS^X was applied to the log-transformed data. (For an introduction to MANOVA, multivariate analysis of variance, see Morrison, 1976, Chapters 4 and 5.) The factorial design is **lines × parachutes**, with **lines** regarded as a random effect and **parachutes** as a fixed effect. The **within-lines** factor is **position**, representing repeated measurements on each line. The results of the repeated-measures analysis, based on the log-normal distribution for breaking strength, are summarized as follows. The mean breaking strength per line varies strongly between parachutes ($p < 0.0005$) as expected, but not between lines ($p = 0.881$). The MANOVA shows that the interactions *parachute × position* and *line × position* are weak, but that the main effect *line* is very evident ($p < 0.0005$). Further, partitioning the *position* effect into orthogonal polynomial contrasts yields only a linear component ($p < 0.0005$), the quadratic and higher order terms not showing strongly. Incidentally, tests on the covariance matrix of the multivariate observations show only small departure from sphericity (e.g. Crowder and Hand, 1990, section 3.6). This

Table 7.1 Breaking strengths of parachute rigging lines

Position							
1	2	3	4	5	6	line	parachute
1104	1236	1215	1208	1201	1181	1	1
1208	1201	1215	1299	1229	1236	2	1
1118	1146	1194	1188	1215	1215	3	1
1167	1201	1215	1236	1194	1264	4	1
1236	1181	1229	1215	1208	1236	5	1
1146	1188	1181	1188	1174	1160	6	1
1174	1181	1181	1194	1194	1160	7	1
1208	1160	1208	1194	1222	1201	8	1
1250	1243	1236	1306	1264	1278	1	2
1264	1222	1250	1222	1201	1285	2	2
1313	1292	1215	1340	1264	1167	3	2
1222	1264	1215	1167	1208	1271	4	2
1285	1167	1285	1236	1271	1299	5	2
1201	1201	1208	1222	1250	1243	6	2
1243	1257	1299	1243	1306	1306	7	2
1264	1313	1236	1299	1250	1174	8	2
938	1076	951	958	944	965	1	3
861	833	889	938	903	861	2	3
938	965	903	910	986	1042	3	3
1000	1000	1014	944	1063	1042	4	3
944	924	944	972	938	847	5	3
889	847	924	938	944	1000	6	3
917	993	1000	882	1014	944	7	3
813	924	931	938	993	993	8	3
1215	1146	1215	1181	1257	1194	1	4
1111	1146	1188	1132	1146	1188	2	4
1111	1250	1208	1243	1285	1215	3	4
1125	1250	1188	1264	1229	1132	4	4
1160	1125	1181	1208	1264	1236	5	4
1208	1125	1229	1111	1208	1299	6	4
1229	1167	1167	1250	1236	1215	7	4
1181	1208	1139	1201	1229	1250	8	4
1215	1264	1243	1306	1215	1313	1	5
1215	1257	1257	1264	1250	1299	2	5
1222	1167	1250	1264	1243	1271	3	5
1104	1222	1208	1208	1257	1222	4	5
1069	1236	1236	1299	1257	1264	5	5
1174	1278	1257	1271	1292	1278	6	5
1222	1194	1236	1236	1250	1306	7	5
1215	1243	1278	1125	1257	1257	8	5

Table 7.1 (*cont.*)

Position 1	2	3	4	5	6	line	parachute
931	972	826	938	951	1014	1	6
924	917	1042	986	986	1007	2	6
958	917	944	931	854	1021	3	6
917	944	938	965	965	1028	4	6
903	882	944	951	938	972	5	6
875	896	972	944	1028	979	6	6
924	924	896	951	938	1042	7	6
951	944	951	958	993	1007	8	6

justifies a univariate mixed-model ANOVA of the data which just confirms the aforementioned findings.

To summarize, the parachutes differ strongly in breaking strengths but the different lines on a parachute do not. Also, there is a (linear) decrease in log-strength as one moves from position 8 down to 1, this probably being due to a higher degree of handling during packing at the lower positions.

Example 7.2

Table 7.2 contains three sets of data on fibre failure strengths. The four values in each row give the breaking strengths of fibre sections of lengths 5, 12, 30 and 75 mm. The four were cut from the same fibre and inspection of the data suggests appreciable variation in strength between fibres. The values are right-censored at 4.0 and a zero indicates accidental breakage prior to testing; the zeros have been treated as missing data.

In view of the apparent heterogeneity between fibres a model allowing individual random levels would seem appropriate and the multivariate Burr described in section 7.2 will be tried. The joint survivor function for the four breaking strengths $t=(t_1, t_2, t_3, t_4)$ is $S(t)=(1+s)^{-v}$, where $s=\sum \xi_j t_j^{\eta_j}$ and the summation is over $j=1, \ldots, 4$. An interpretation of the parameters may be based on the facts that for this model $\log t_j$ has mean $-\eta_j^{-1}\{\log \xi_j+\gamma+\psi(v)\}$ and variance $\eta_j^{-2}\{\pi^2/6+\psi'(v)\}$; here γ is Euler's constant and $\psi(v)=\partial \log \Gamma(v)/\partial v$ is the digamma function. Thus $\log \xi_j$ is a location parameter for $\log t_j$ and η_j is a scale parameter. Also, $\log t_j$ and $\log t_k$ have correlation $[1+\pi^2/\{6\psi'(v)\}]^{-1}$, which identifies v as an association parameter.

The log-likelihood function for an individual vector observation is constructed as follows. Suppose, as an example, that the observation is $(t_1, 0, t_3, 4.0)$; thus t_1 and t_3 are actually recorded, t_2 is missing, and t_4 is censored. The

Table 7.2 Fibre failure strengths: three data sets, fibre sections of different lengths

Data set 1				Data set 2				Data set 3			
Fibre length (mm)				Fibre length (mm)				Fibre length (mm)			
No. 5	12	30	75	No. 5	12	30	75	No. 5	12	30	75
1 3.30	3.32	2.39	2.08	1 4.00	4.00	2.80	2.29	1 4.00	4.00	2.39	2.22
2 4.00	3.67	2.49	2.06	2 4.00	4.00	2.71	2.28	2 3.98	3.20	2.30	2.18
3 4.00	4.00	3.16	2.05	3 3.94	3.86	2.40	2.25	3 4.00	4.00	2.86	2.52
4 3.64	2.41	2.20	1.80	4 4.00	3.35	2.31	2.10	4 4.00	4.00	2.96	2.49
5 2.73	2.24	1.91	1.68	5 3.80	3.01	2.29	2.17	5 4.00	4.00	2.78	2.30
6 4.00	4.00	2.74	2.22	6 3.85	2.28	2.19	1.92	6 4.00	3.82	2.62	2.28
7 3.29	3.08	2.44	2.37	7 4.00	4.00	2.98	2.40	7 4.00	3.40	2.51	2.24
8 3.55	2.35	2.38	2.37	8 3.53	2.20	2.16	1.63	8 4.00	4.00	2.56	2.25
9 3.03	2.26	1.64	2.03	9 3.72	2.28	2.20	1.96	9 4.00	3.65	2.56	2.26
10 4.00	4.00	2.98	2.39	10 3.90	3.01	2.25	2.08	10 4.00	4.00	2.73	2.38
11 4.00	4.00	2.99	2.30	11 4.00	3.08	2.36	2.24	11 4.00	4.00	2.80	2.48
12 4.00	3.03	2.80	2.30	12 4.00	3.67	2.49	2.20	12 4.00	4.00	3.00	2.41
13 3.01	3.17	2.41	2.07	13 4.00	4.00	2.86	2.15	13 3.54	3.09	2.24	1.73
14 4.00	2.91	2.18	1.83	14 4.00	4.00	2.64	2.32	14 3.40	2.32	2.18	1.71
15 4.00	3.87	2.24	2.09	15 4.00	4.00	2.80	2.40	15 3.28	2.18	2.12	1.72
16 4.00	3.82	2.59	2.48	16 3.72	3.32	2.21	2.10	16 3.02	2.14	2.04	1.52
17 4.00	4.00	2.40	2.22	17 3.82	3.40	2.28	2.11	17 3.06	2.14	2.09	1.52
18 3.48	2.14	2.35	2.05	18 3.92	3.38	2.28	2.54	18 3.20	2.30	2.08	1.58
19 3.05	2.96	1.91	2.20	19 4.00	3.39	2.36	2.21	19 3.96	3.18	2.30	2.19
20 3.60	2.92	2.42	2.09	20 4.00	3.08	2.49	2.23	20 3.73	3.22	2.25	1.89
21 4.00	4.00	2.86	2.13	21 3.28	2.29	2.18	1.83	21 3.70	3.28	2.17	2.04
22 4.00	3.03	2.53	2.31	22 3.35	2.25	2.18	1.71	22 4.00	3.20	2.36	2.40
23 3.50	3.46	2.56	2.13	23 3.71	2.30	2.19	2.05	23 3.83	3.16	2.28	2.09
24 4.00	4.00	2.63	2.16	24 3.98	3.30	2.23	2.03	24 3.92	3.25	2.29	2.10
25 4.00	4.00	2.75	2.17	25 3.70	3.00	2.22	2.02	25 4.00	2.16	2.22	2.00
26 4.00	3.64	2.88	2.43	26 4.00	3.30	2.42	2.22	26 4.00	3.22	2.28	2.21
27 4.00	3.20	2.52	2.35	27 4.00	4.00	2.50	2.32	27 4.00	3.28	2.34	2.81
28 2.65	2.01	1.87	2.12	28 4.00	4.00	3.12	2.53	28 3.68	2.51	2.24	1.63
29 4.00	3.85	3.12	2.53	29 3.68	2.28	2.20	2.09	29 3.15	2.48	2.37	1.80
30 4.00	3.35	2.78	2.36	30 3.10	2.18	2.10	1.73	30 4.00	3.36	2.40	2.26
31 3.35	2.91	2.50	2.07	31 3.30	2.29	2.23	1.58	31 4.00	3.42	2.41	2.22
32 3.62	3.31	2.50	2.08	32 3.23	2.28	2.08	1.53	32 4.00	3.40	2.38	2.24
33 4.00	3.35	2.41	2.37	33 2.94	2.14	2.06	1.53	33 4.00	3.52	2.36	2.20
34 3.06	2.49	2.09	2.21					34 3.82	3.72	2.40	2.23
35 4.00	2.67	2.40	2.28								
36 3.23	2.27	1.92	2.12								
37 4.00	4.00	3.47	2.24								
38 3.75	2.48	2.48	2.07								
39 3.33	2.23	2.33	2.13								
40 3.47	2.51	0.0	1.76								
41 3.70	2.31	0.0	2.06								

Table 7.2 (*cont.*)

Data set 1 Fibre length (mm) No. 5		12	30	75	Data set 2 Fibre length (mm) No. 5		12	30	75	Data set 3 Fibre length (mm) No. 5		12	30	75
42	3.77	2.26	0.0	2.20										
43	0.0	2.37	0.0	0.0										
44	0.0	2.39	0.0	0.0										
45	0.0	2.41	0.0	0.0										

log-likelihood function for this observation is then

$$\frac{\partial^2}{\partial t_1 \partial t_3}(1+s)^{-\nu}$$

where

$$s = \xi_1 t_1^{\eta_1} + \xi_3 t_3^{\eta_3} + \xi_4(4.0)^{\eta_4}.$$

Note that t_4 has been set at 4.0 in the expression for s, and t_2 has been set at zero; in effect, t_2 is right-censored at 0. The differentiations are performed only for the completely observed times. The log-likelihood for the whole sample is then just the sum of all such contributions from the individual observations.

The maximized log-likelihoods resulting from fitting the model separately to the three data sets are -71.975, -26.313 and -36.900; that resulting from fitting the three data sets together is -166.523. The difference, $166.523 - (71.975 + 26.313 + 36.900) = 31.335$, is due to fitting only one set of nine parameters (four ξs, four ηs and ν) to all three data sets rather than three sets of nine parameters. The log-likelihood ratio test of equality of parameters between the data sets is then performed by referring twice the difference, 62.670, to the χ^2 distribution with $18 = 27 - 9$ degrees of freedom. From tables $p < 0.001$, strongly indicating differences between the data sets.

Table 7.3 gives the maximum likelihood estimates of the parameters. There $\beta_j = \log \xi_j$ is used because

1. this parameter transformation has been found to improve the convergence properties of the iterative maximization procedure used for fitting the model, and
2. for data with explanatory variables x, there is the natural generalization to a loglinear model $\log \xi_j = x^T \beta_j$.

Comparison of the β-values in Table 7.3 between data sets is not rewarding. In a first attempt at standardization the values $-\beta_j/\eta_j$ have been tabulated. This

Table 7.3 Maximum likelihood estimates of parameters of multivariate Burr distributions fitted to fibre-strength data

	Data Set 1	Data Set 2	Data Set 3
β_1	-7.14	-14.27	-12.75
β_2	-6.40	-7.46	-8.23
β_3	-11.56	-20.54	-20.32
β_4	-16.44	-13.05	-10.01
η_1	5.26	11.78	10.45
η_2	6.03	8.81	9.13
η_3	13.64	27.20	26.62
η_4	21.77	21.36	17.27
v	0.684	0.310	0.303
$-\beta_1/\eta_1$	1.36	1.21	1.22
$-\beta_2/\eta_2$	1.06	0.85	0.90
$-\beta_3/\eta_3$	0.85	0.75	0.76
$-\beta_4/\eta_4$	0.76	0.61	0.58
$-\eta_1^{-1}\{\beta_1+\psi(v)\}$	1.60	1.50	1.55
$-\eta_2^{-1}\{\beta_2+\psi(v)\}$	1.27	1.23	1.28
$-\eta_3^{-1}\{\beta_3+\psi(v)\}$	0.94	0.88	0.89
$-\eta_4^{-1}\{\beta_4+\psi(v)\}$	0.81	0.77	0.78
$\eta_1^{-2}\{\pi^2/6+\psi'(v)\}$	0.17	0.09	0.13
$\eta_2^{-2}\{\pi^2/6+\psi'(v)\}$	0.13	0.17	0.16
$\eta_3^{-2}\{\pi^2/6+\psi'(v)\}$	0.02	0.18	0.02
$\eta_4^{-2}\{\pi^2/6+\psi'(v)\}$	0.01	0.03	0.05

corresponds to replacing $s=\sum \xi_j t_j^{\eta_j}$ in the survivor function by $s=\sum (t_j/e^{-\beta_j/\eta_j})^{\eta_j}$, so that $e^{-\beta_j/\eta_j}$ is a scale parameter for t_j. These figures are evidently rather more comparable over different data sets. The further standardization $-\eta_j^{-1}\{\beta_j+\psi(v)\}$ is also tabulated, this being suggested by the expected value of $\log t_j$ quoted above. With the influences of η and v removed from β in this way it is seen that the data sets are fairly homogeneous in this respect. In this manner the η_js in Table 7.3 differ markedly between data sets. Even when the standardized version $\eta_j\{\pi^2/6+\psi'(v)\}$, suggested by the variance formula for $\log t_j$, is applied, there is still considerable disagreement. It appears that a major difference between data sets rests in the v-estimates though these are not well defined by the data, having standard errors of about 0.3. (The values used above for $\psi(v)$ were -1.27, -3.38 and -3.47 for the three data sets, and those for $\psi'(v)$ were 2.94, 11.53 and 12.02).

A basic model for breakage of a single strand of material stems from the analogy with the weakest link in a chain of elements. One consequence is that the survivor function S_l for breakage of fibres of length l under load x should be such that $\{S_l(x)\}^{1/l}$ is independent of l. A Weibull distribution then has form

$S_l(x) = \exp(-l\lambda x^\eta)$, and this results in the multivariate Burr form $S(t) = (1+s)^{-\nu}$ with $s = l\xi\sum t_j^\eta$. In terms of the parameters used above, the weakest link model requires that $\eta_j = \eta$ and $l_j^{-1}e^{-\beta_j} = \xi$ for $j = 1, \ldots, 4$. These relations clearly fail for the estimates in Table 7.3. For instance, $\hat{\eta}_1$ and $\hat{\eta}_2$ are substantially smaller than $\hat{\eta}_3$ and $\hat{\eta}_4$, consistently over data sets, and the $\hat{\beta}_j$ do not decrease steadily with j.

Goodness-of-fit of the model may be assessed by plotting residuals as now illustrated for data set 1. The uniform residuals may be calculated from Rosenblatt's (1952) extension of the probability integral transform. Let $u_1 = S_1(t_1)$ and $u_j = S_j(t_j|t_1, \ldots, t_{j-1})$; here S_1 denotes the marginal survivor function of t_1 and $S_j(t_j|t_1, \ldots, t_{j-1})$ the conditional survivor function of t_j given the previous ts. Then $u_1, \ldots, u_p$ are independent, standard uniform variates. The application of the transform to the current data is hindered by the high degree of censoring in t_1 and, to a lesser extent, in t_2. Figure 7.1 shows the us corresponding to t_4 (symbol $+$) and $t_3|t_4$ (symbol $\times$) plotted against expected uniform order statistics $\{i/(n+1): i = 1, \ldots, n\}$; diagrams (a), (b) and (c) correspond to data sets 1, 2 and 3 respectively. There seems to be significant departure from the $45°$ line in (b) and (c), thus casting doubt on the assumed distribution for the data. Such failure may not affect predictions made from the fitted model too seriously but in an extended study it would be more satisfactory to find a distributional model which fitted the data well and had some physical credibility.

7.4 COMPETING RISKS

The classical case of incomplete observation of T is known as 'Competing Risks' and was described in section 7.1. In this situation only $U = \min(T_1, \ldots, T_p)$ and C, where $U = T_C$, are observable; hence the identifiable aspect of the model is the so-called sub-survivor function $P(c, u) = \Pr[C = c, U \geq u]$, representing the joint distribution of C and U. We give here an outline of the theory slanted towards reliability application. Standard references include Cox (1959), Tsiatis (1975) and Gail (1975).

If the joint survivor function of T has known form $S(t)$, then the survivor function of U can be calculated as $S_U(u) = S(u1_p)$, where $1_p = (1, \ldots, 1)$. Also, $\Pr[C = c] = P(c, 0) = p_c$, say, and $S_U(u) = \sum_{c=1}^{p} P(c, u)$ give the marginal distributions of C and U in terms of $P(c, u)$. The sub-density function corresponding to $P(c, u)$ is given by

$$p(c, u) = [-\partial S(t)/\partial t_c]_{u1_p}, \quad P(c, u) = \int_u^\infty p(c, t)\, dt,$$

the notation $[\ ]_{u1_p}$ indicating that the enclosed function is to be evaluated at

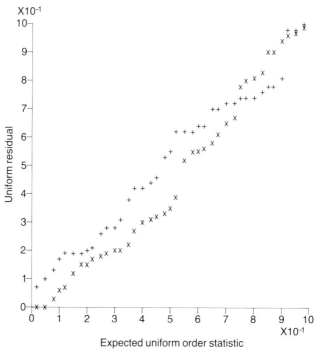

(a)

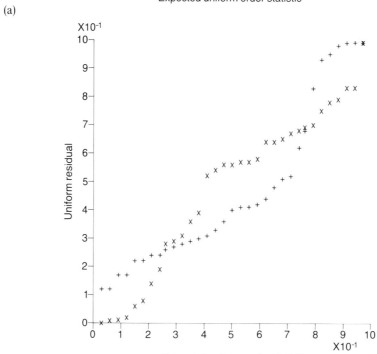

(b)

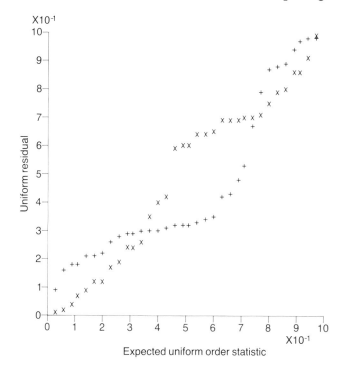

(c)

Figure 7.1(a)(b)(c) Residuals from fitted multivariate Burr model.

$t = u1_p$. The marginal density of U is

$$p(u) = \partial S_U(u)/\partial u = \sum_{c=1}^{p} p(c, u),$$

and related conditional probabilities are $\Pr[\text{time } u|\text{type } c] = p(c, u)/p_c$ and $\Pr[\text{type } c|\text{time } u] = p(c, u)/p(u)$. Note that $\Pr[\text{time } u|\text{type } c]$ is the conditional density of T_c given $T_c = \min(T_j)$. This is not generally equal to the marginal density $-\partial S_c(t)/\partial t$ of T_c, where $S_c(t)$ is the marginal survivor function of T_c given by $S_c(t_c) = S(0 \ldots t_c \ldots 0)$. 'Independent risks' obtains if $S(t) = \prod_{j=1}^{p} S_j(t_j)$, i.e. if the T_j are independent.

Various hazard functions are associated with the set up. The overall hazard rate from all causes is $h(u) = -\partial \log S_U(u)/\partial u$, and that for type c failure, in the presence of all risks 1 to p, is

$$h_c(u) = p(c, u)/S_U(u) = [-\partial \log S(t)/\partial t_c]_{u1_p};$$

thus $h(u) = \sum_{c=1}^{p} h_c(u)$, $h_c(u)$ being a special case of (7.1) with $t = u1_p$. The relative risk of type c failure at time u is $h_c(u)/h(u)$. If this is independent of u then 'proportional hazards' are said to obtain, the term being used in a different sense to that in Chapter 4. It can be shown that proportional hazards obtains if and only if $h_j(t)/h_k(t)$ is independent of t for all j and k, and also if and only if the time and cause of failure are independent; in this case $h_j(t) = p_j h(u)$. Thus, for example, failure during some particular period does not make it any more or less likely to be from cause 1 than in any other period.

In certain special circumstances it may be possible to study the effect of failure type c alone (i.e. with other types absent), and then the marginal hazard rate $r_c(u) = -\partial \log S_c(u)/\partial u$ is relevant. However, it is often the case that elimination of the other types of failure changes the circumstances so that it is not valid to assume that the isolated T_c-distribution is the same as the former marginal T_c-distribution.

In the special case of independent risks (i) $h_j(u) = r_j(u)$, and (ii) proportional hazards obtains if and only if $S_j(u) = S(u)^{p_j}$. Note that (ii) is reminiscent of the 'proportional hazards' property in Chapter 4. Also, (iii) the sets of functions $\{P(j, u): j = 1, \ldots, p\}$ and $\{S_j(u): j = 1, \ldots, p\}$ determine each other. This means that the observable distributions $P(c, u)$ and the marginal distributions $S_j(t)$ give equivalent descriptions of the set-up. The identity (i) is vital in the proof of (iii) and it is of interest that this can hold even when the risks are not independent. For such models the conditional hazard rate for T_j, given no failures so far, is equal to the unconditional one. Cox and Oakes (1984, section 9.2) note this as a hypothetical possibility, and Williams and Lagakos (1977, section 3) given an example in which $p = 2$, T_1 and T_2 are dependent, and $h_1(u) = r_1(u)$.

Suppose that the set of sub-survivor functions $P(j, u)$ is given for some model with dependent risks. Then there is a 'proxy model' with independent risks yielding identical $P(j, u)$. This is a non-identifiability result which says that one cannot establish from observations of (C, U) alone the form of $S(t)$, nor, in particular, whether the failure times are independent or not. This aspect has brought the method of fitting an essentially untestable form of $S(t)$ as a basis for the $P(c, u)$ into some disrepute.

Example 7.3

Gumbel's first bivariate exponential distribution in section 7.2 has $S(t) = \exp(-t_1 - t_2 - \delta t_1 t_2)$. In this case $p(c, u) = (1 + \delta u) \exp(-2u - \delta u^2)$ for $c = 1, 2$ and the alias model has $S_j^*(u) = \exp(-u - \delta u^2/2)$. Hence, for example, predictions about T_j based on $S_j(u) = \exp(-u)$ differ from those based on $S_j^*(u)$, and one can never know purely from data which should be used.

The non-identifiability only causes problems when one tries to make inferences about aspects outside the scope of the $P(c, u)$ functions. The example illustrates the impossibility of making sensible predictions about T_j, i.e. from its marginal distribution, solely from the given $p(c, u)$. However, predictions about (C, U) are perfectly respectable. If the T_j are known to be independent, e.g. being the lifetimes of independently operating components in a system, then the problem disappears. In that case result (i) above implies that the joint distribution, represented by $S(t)$, is completely identified once the $P(c, u)$ functions are given.

When the T_j are not known to be independent the non-identifiability problem remains. An obvious suggestion is to use a plausible parametric model $S(t)$ only as a basis for the $P(c, u)$ and then refrain from making any invalid predictions. The assumed form $S(t)$ will provide a likelihood function and so the model may be fitted, say by maximum likelihood. An alternative suggestion (Prentice *et al.* 1978) is to model the $P(c, u)$s, or equivalently the $h_j(u)$s, directly.

Various goodness-of-fit tests may be applied to compare estimated quantities with their observed counterparts, e.g. $p_j = P(j, 0)$ (the proportion of type j failures), $P(j, u)/p_j$ (the survivor function of type j failures), $\sum_j p(j, u)$ (the survivor function of all failures). Further discussion, including a non-parametric approach, may be found in Lawless (1982, section 10.1.2).

Example 7.4
Mendenhall and Hader (1958) present an analysis for competing risks and illustrate it with the data reproduced here in Table 7.4.

Briefly, these are times to failure of radio transmitters with two types of failure and censoring at 630 hours. The authors fit, by maximum likelihood, a model representing a mixture of two exponential distributions, one for each failure type, in proportion $p: 1 - p$. In the present notation their model has $S_U(u) = p_1 S_1(u) + p_2 S_2(u)$, where $S_j(u) = \exp(-u/\alpha_j)$ for $j = 1, 2$ and $p_1 = p$, $p_2 = 1 - p$; also, $p(j, u) = (p_j/\alpha_j) \exp(-u/\alpha_j)$ from which the likelihood function can be constructed. Such a mixture model is very natural when there are two distinct types of system, the type being identifiable only at failure. It seems less compelling, though certainly not inadmissible, for a single population subjected to competing causes of failure.

A Weibull mixture loglinear model, as described in section 7.2, will be applied with joint survivor function $S(t) = \exp\{\kappa^v - (\kappa + s)^v\}$ where $s = \xi_1 t_1^{\eta_1} + \xi_2 t_2^{\eta_2}$ and $\log \xi_j = \beta_j$ ($j = 1, 2$). For this model $S_U(u) = \exp\{\kappa^v - (\kappa + s_u)^v\}$ with $s_u = \xi_1 u^{\eta_1} + \xi_2 u^{\eta_2}$, and the sub-densities are

$$p(j, u) = v \xi_j \eta_j u^{\eta_j - 1} (\kappa + s_u)^{v - 1} S_U(u).$$

In the alternative interpretation of Prentice *et al.* (1978) the model is specified by the hazard functions $h_j(u) = v \xi_j \eta_j u^{\eta_j - 1} (\kappa + s_u)^{v - 1}$. The log-likelihood function for data with n_j observed failures of type j ($j = 1, 2$) and n_0 right-censored

Table 7.4 Failure times of radio transmitters

Type 1 failures, n = 107

368	136	512	136	472	96	144	112	104	104
344	246	72	80	312	24	128	304	16	320
560	168	120	616	24	176	16	24	32	232
32	112	56	184	40	256	160	456	48	24
200	72	168	288	112	80	584	368	272	208
144	208	114	480	114	392	120	48	104	272
64	112	96	64	360	136	168	176	256	112
104	272	320	8	440	224	280	8	56	216
120	256	104	104	8	304	240	88	248	472
304	88	200	392	168	72	40	88	176	216
152	184	400	424	88	152	184			

Type 2 failures, n = 218

16	224	16	80	128	168	144	176	176	568
392	576	128	56	112	160	384	600	40	416
408	384	256	246	184	440	64	104	168	408
304	16	72	8	88	160	48	168	80	512
208	194	136	224	32	504	40	120	320	48
256	216	168	184	144	224	488	304	40	160
488	120	208	32	112	288	336	256	40	296
60	208	440	104	528	384	264	360	80	96
360	232	40	112	120	32	56	280	104	168
56	72	64	40	480	152	48	56	328	192
168	168	114	280	128	416	392	160	144	208
96	536	400	80	40	112	160	104	224	336
616	224	40	32	192	126	392	288	248	120
328	464	448	616	168	112	448	296	328	56
80	72	56	608	144	408	16	560	144	612
80	16	424	264	256	528	56	256	112	544
552	72	184	240	128	40	600	96	24	184
272	152	328	480	96	296	592	400	8	280
72	168	40	152	488	480	40	576	392	552
112	288	168	352	160	272	320	80	296	248
184	264	96	224	592	176	256	344	360	184
152	208	160	176	72	584	144	176		

failure times is

$$\sum_1 \log p(1, t_{1i}) + \sum_2 \log p(2, t_{2i}) + \sum_0 \log S_U(t_{0i});$$

here $\sum_j$ denotes summation over, and the t_{ji} denote the times of, failures of type j. Maximization of this over the parameters $(\beta_1, \beta_2, \eta_1, \eta_2, \nu, \kappa)$ for the data of Table 7.4 yields results summarized in Table 7.5.

Table 7.5 Maximum likelihood estimates for Weibull mixture models fitted to competing risks data

Model constraints	β_1	β_2	ϕ_1	ϕ_2	v	κ	$-log\text{-}likelihood$
none	-1.52	-1.04	1.45	1.68	0.43	0.52	880.45773
$\kappa=0$	-12.98	-12.37	10.43	10.65	0.11	0	885.18801
$\phi_1=\phi_2$	-1.63	-0.92	1.57	1.57	0.44	0.59	882.05011
(s.e.)	(2.4)	(2.4)	(0.26)	(0.26)	(0.20)	(2.2)	

Let us first decide whether the parameter κ can be dispensed with, i.e. set to its null value zero. The maximized log-likelihood under the constraint $\kappa=0$ is -885.18801, and this differs from the unconstrained maximum -880.45773 by 4.73028. An ordinary likelihood ratio test would take twice the difference, 9.46056, as a χ_1^2 variate whence $p<0.005$, implying very little support for the null value. There is a slight complication here in that zero is a boundary point for κ since the probability model is only defined on $\kappa \geq 0$. In this case the asymptotic distribution theory should properly be modified (Self and Liang, 1987; Crowder, 1990) but this diversion will be ignored here. In Crowder (1989) κ is given the interpretation of a preliminary conditioning parameter in the sense than an item can only be eligible for the sample after passing some pre-test. This accords with manufacturing quality control practice for items such as radios. Figure 7.2 shows the profile

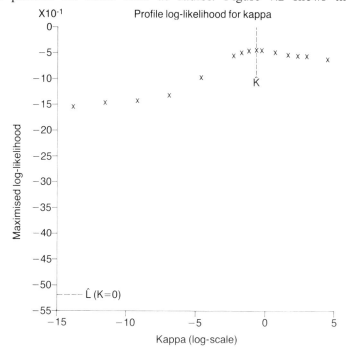

Figure 7.2 Profile log-likelihood for κ.

log-likelihood for κ, i.e. the values of the log-likelihood with κ fixed and the other parameters varied to maximize it.

The curve is very flat so κ is not at all sharply identified by the data. For instance, a conventional likelihood ratio 90% confidence interval for κ would be defined as the range of values over which the profile log-likelihood lies above the level -881.81073; this is 1.353 below the unconstrained maximum, and $2 \times 1.353 = 2.706$, the upper 10% point of χ_1^2. The level is below all the points in Figure 7.2 so the interval includes the whole κ-range shown. In fact, numerical experiment extends the range included to $(10^{-6}, 150)$, and it is not impossible that the whole line $(0, \infty)$ is covered, exhibiting a well known drawback of confidence intervals. However, in view of the previous test, the value 0 is not included; Self and Liang (1987, section 4) comment upon such discontinuity at boundary points.

Another parametric question, suggested by the unconstrained fit, is whether η_1 and η_2 are equal. The maximized log-likelihood under the constraint $\eta_1 = \eta_2$ is -882.05011, yielding $\chi_1^2 = 3.18$ ($p > 0.05$), so this is not strongly contradicted by the data.

The general goodness-of-fit of the model with $\eta_1 = \eta_2$ ($= \eta$) can be assessed by examining residuals as follows. The survivor function for failures of type j is $\Pr(U > u | C = j) = P(j, u)/P(j, 0)$. Here $P(j, u) = p_j S_U(u)$ with $p_j = \xi_j/\xi_+$, $S_U(u) = \exp\{\kappa^\nu - (\kappa + s_u)^\nu\}$, $s_u = \xi_+ u^\eta$, and $\xi_+ = \xi_1 + \xi_2$. Hence, the conditional survivor function in question is $S_U(u)$, the same for $j = 1$ and 2; this is an example of proportional hazards. Applying the probability integral transform, the quantities $S_U(t_{ji})$ ($i = 1, \ldots, m_j$) for $j = 1, 2$ should resemble a random sample from U(0, 1), the uniform distribution on [0, 1]. An adjustment has to be made for the unobserved times to failure of type j, i.e. those among the censored observations. Thus m_j comprises n_j, the number of observed times, plus a proportion of n_0, the number of censored times. Denoting the ordered observed times by $t_{j(1)} < t_{j(2)} < \cdots < t_{j(n_j)}$, the quantities $S_U(t_{j(i)})$ ($i = 1, \ldots, n_j$) should resemble the first n_j order statistics in a sample of size m_j from U(0, 1); the latter have expected values $i/(m_j + 1)$. Figure 7.3 gives plots in which the points are estimates of the pairs $\{i/(m_j + 1), S_U(t_{j(i)})\}$.

The S_U function just has the maximum likelihood estimates inserted, and m_j is estimated as $n_j + p_j n_0$. The latter arises from the fact that the expected proportion of failures of type j beyond the censoring time is

$$\Pr(C = j | U > u_0) = P(j, u_0)/S_U(u_0) = \xi_j/\xi_+ = p_j.$$

From Table 7.5, $\hat{\xi}_1 = \exp(\hat{\beta}_1) = 0.196$ and $\hat{\xi}_2 = \exp(\hat{\beta}_2) = 0.399$, so $\hat{p}_1 = 0.329 = 1 - \hat{p}_2$; $\hat{p}_1$ agrees well with $n_1/(n_1 + n_2)$. There seems to be reasonable adherence in Figure 7.3 to the target 45° line for both failure types.

In conclusion, the data appear to be in accordance with the mixed Weibull model fitted, judging by the residual plots, though, of course, this does not

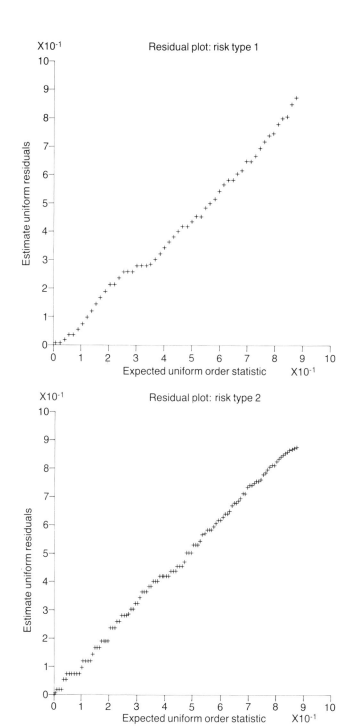

(a)

(b)

Figure 7.3 Residual plots for Mendenhall-Hader data.

mean that the model is necessarily 'correct'. The pre-testing parameter, κ in the model, seems to be non-null, and the shape parameters for the distributions of the two failure types, η_1 and η_2, are not obviously unequal. The fitted model may be used to make predictions such as the likely proportion of type-1 failures and the proportion of transmitters likely to fail within a given time.

8

Repairable systems

8.1 INTRODUCTION

A system may be regarded as a collection of two or more components which is required to perform one or more functions. A repairable system is a system which, after it has failed to perform properly, can be restored to satisfactory performance by any method except replacement of the entire system. In practice many systems are of this type. Examples featured in this chapter are: air-conditioning equipment in an aircraft; a system of generators in a marine vessel, main propulsion diesel engines of USS Grampus and USS Halfbeak; and some software during the production and testing stages.

Note that the above definition allows the possibility of repair without replacing any components. For example, a well-aimed blow with a hammer may be sufficient to effect a repair.

The aim of this chapter is to introduce some simple statistical methods for dealing with repairable systems reliability data.

8.2 FRAMEWORK

Consider a repairable system. Let $T_1, T_2, T_3, \ldots$ be the times of failure of the system and let $X_i (i = 1, 2, 3, \ldots)$ be the time between failure $i-1$ and failure i, where T_0 is taken to be zero. The T_i and X_i are random variables. Let t_i and x_i be the corresponding observed values and let $N(t)$ be the number of failures in the time interval $(0, t]$.

Of particular interest in reliability is the behaviour of the X_i. Ascher and Feingold (1984) speak of happy and sad systems in which the times between failures are tending to increase and decrease respectively. Thus the detection and estimation of trends in the X_i is a major concern. For example, it may be vital to weed out sad systems which might disrupt a production process.

Even if there is no long-term trend in the X_i, there may be other structures of practical importance. For example, suppose the repairs are sometimes inadequate. Then there will be runs of short times between failures caused by the poor repairs. Once the necessary repairs have been carried out satisfactorily, there may well be a relatively long time to the next failure, and so on. Thus, the occurrence of dependence or autocorrelation betwen the X_i or of some other form of localized behaviour may be of interest.

Note that in component reliability, where one generally has times to first failure for several different components of the same type, the assumption of independent and identically distributed (IID) lifetimes may be a natural one. However, in repairable systems one generally has times of *successive* failures of a single system and it is the departures from the IID assumption that are important. Thus, it is not surprising that the statistical methods required for repairable systems differ somewhat from those needed in component reliability.

In the above discussion we have used time in its most general sense. In any particular application, time needs to be defined carefully and appropriately. For a continuously operating refrigerator in reasonably constant conditions, real time may be an appropriate measure for investigating reliability of the system. In such cases we shall assume that any restoration to satisfactory performance after failure is done instantaneously. More commonly, however, operating time may be more appropriate than real time. On the other hand, some non-negative quantity other than time itself may be suitable. For example, the mileages of a motor vehicle between breakdowns may be more useful for assessing reliability than real time and easier to record than operating time.

An inappropriate choice of the measure of time to be used may make it impossible to obtain useful results on reliability. Thus, it is important that the right numerical data should be collected from the start of the period of observation. Failure to do so may have serious consequences.

8.3 ROCOF

The rate of occurrence of failures (ROCOF) is defined by

$$v(t) = \frac{\mathrm{d}}{\mathrm{d}t} E\{N(t)\}.$$

A happy (sad) system will have decreasing (increasing) ROCOF. However, note that just because a system is happy (sad) does not necessarily mean that, in practical terms, it is satisfactory (unsatisfactory). For example, a system with very low ROCOF may be perfectly satisfactory for its planned useful life even though its ROCOF is increasing.

Note that the ROCOF, which is sometimes called the failure rate, should not be confused with the hazard function, which is also sometimes known as the failure rate. See Ascher and Feingold (1984, Chapter 8) for a discussion of poor terminology. It is possible for each of the X_i for a system to have non-decreasing hazard and for the system to have a decreasing ROCOF.

A natural estimator of $v(t)$ is $\hat{v}(t)$, given by

$$\hat{v}(t) = \frac{\text{number of failures in } (t, t+\delta t]}{\delta t} \tag{8.1}$$

for some suitable δt. The choice of δt is arbitrary, but, as with choosing the interval width for a histogram, the idea is to highlight the main features of the data by smoothing out the 'noise'.

8.4 SIMPLE STATISTICAL METHODS

Method (a)
A simple but informative graph comprises the cumulative number of failures versus the cumulative time; that is i versus t_i. Departures from linearity are indicative of the fact that the X_i are not IID. In particular this plot is useful for detecting the presence of a trend. To gain an indication as to the form of the trend the estimated ROCOF (8.1) may be calculated and/or plotted against time. See also section 8.8.

Example 8.1
The 23 observations in Table 8.1 are the intervals (in operating hours) between failures of the air-conditioning equipment in a Boeing 720 aircraft. A glance at these data suggests that towards the end of the period of observation failures were tending to occur much more frequently than at the beginning. This impression is confirmed by Figure 8.1 and also by the estimated ROCOF, which is proportional to

$$2 \quad 6 \quad 1 \quad 3 \quad 11,$$

where we have taken δt in (8.1) to be about 440 hours.

Table 8.1 Inter-failure times in operating hours of air-conditioning equipment. Here 413 is the first interval and 34 is the last. Source: Proschan (1963)

413	14	58	37	100	65	9	169
447	184	36	201	118	34	31	18
18	67	57	62	7	22	34	

It would be misleading to analyse these data as IID observations. Indeed, after noticing the main feature of the data it was discovered that there had been a major overhaul of the equipment after the 13th failure, from which point the reliability of the system changed drastically for the worse.

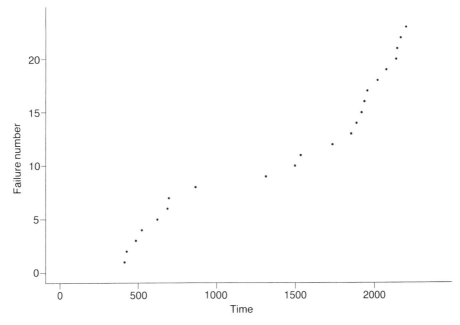

Figure 8.1 Plot of failure number against time in operating hours for the air-conditioning equipment data.

Example 8.2

The data in Table 8.2 from Triner and Phillips (1986) are the number of failures of a system of generators on a marine vessel. Note that we do not have the t_i, only the numbers of failures within non-overlapping time intervals. Even so we can still plot the cumulative number of failures against the upper end-point of each interval. This plot is shown in Figure 8.2. The estimated ROCOF $\hat{v}(t)$ is

Table 8.2 Numbers of failures and estimated ROCOF of a system of generators on a marine vessel

Time interval (years)	Failures	Estimated ROCOF
1.5– 2.5	4	4
2.5– 3.5	5	5
3.5– 4.5	4	4
4.5– 5.5	2	2
5.5– 6.5	4	4
6.5– 7.5	11	11
7.5– 8.5	19	19
8.5– 9.5	10	10
9.5–10.33	14	16.9

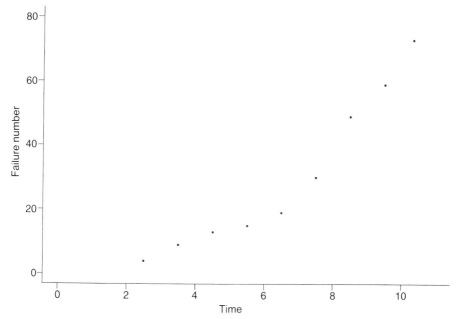

Figure 8.2 Plot of failure number against time in years for the generators data.

also shown in Table 8.2 where we have used the given intervals to calculate $\hat{v}(t)$. Once again, failures seem to be occurring more frequently as time goes on, which is indicative of a sad system.

Example 8.3
The data in Table 8.3 are a subset of the times in operating hours of unscheduled maintenance actions for a diesel engine of USS Grampus, given in Lee (1980). The plot of these values against the cumulative number of failures, shown in Figure 8.3, looks roughly linear, suggesting a roughly constant ROCOF; that is, no apparent trend in the X_i. We shall return to this example later.

Table 8.3 Times in operating hours of unscheduled maintenance actions for a diesel engine of USS Grampus

860	1 258	1 317	1 442	1 897	2 011	2 122	2 439	3 203	3 298
3 902	3 910	4 000	4 247	4 411	4 456	4 517	4 899	4 910	5 676
5 755	6 137	6 221	6 311	6 613	6 975	7 335	8 158	8 498	8 690
9 042	9 330	9 394	9 426	9 872	10 191	11 511	11 575	12 100	12 126
12 368	12 681	12 795	13 399	13 668	13 780	13 877	14 007	14 028	14 035
14 173	14 173	14 449	14 587	14 610	15 070				

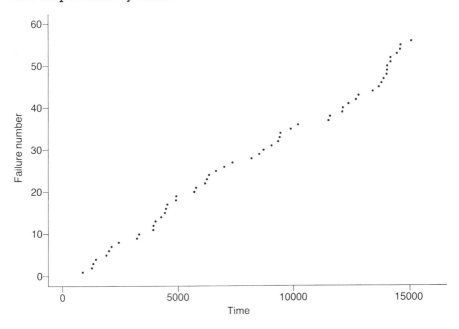

Figure 8.3 Plot of failure number against time in operating hours for the USS Grampus data.

Method (b)
If there is no apparent trend, it may be of interest to investigate the possibility of autocorrelation amongst the X_i. A simple graphical method often used in time series analysis (see, for example, Chatfield, 1980) is to plot X_{i+k} against X_i for various integer values of k, which is known as the lag. Unfortunately, by the very nature of reliability data, which are often highly skewed, such plots in the present context may not be easy to interpret. However, a simple transformation (such as log, square root or cube root) applied to the X_i will often give a plot with roughly normal marginal distribution. After such a transformation one would expect a lag-k plot to have roughly circular contours with the frequency of points decreasing as one moves away from the centre if there is no autocorrelation at lag k.

Another procedure, often used in time series analysis, is to evaluate the lag k sample autocorrelation coefficients r_k for various values of k where

$$r_k = \sum_{i=1}^{n-k} (x_i - \bar{x})(x_{i+k} - \bar{x}) \bigg/ \sum_{i=1}^{n} (x_i - \bar{x})^2,$$

n is the observed number of failures and $\bar{x}$ is the sample mean time between failures. If there is appreciable lag k autocorrelation, we should expect $|r_k|$ to be large. However, interpretation is not entirely straightforward since, even for an entirely random series of failures, the fact that r_k is being evaluated for several values of k means that we should expect the occasional large value amongst the $|r_k|$ just by chance. In time series, the fact that, for a random series, r_k is approximately Normally distributed with zero mean and variance n^{-1} is often used. However, when dealing with highly skewed reliability data the usefulness of this result is somewhat diminished.

Interpretation of the autocorrelation results may be rather difficult in the repairable systems context. Situations typically encountered are that there is a monotonic trend in the X_i, or, because of imperfect repair, a large value of X_i tends to be followed by several small inter-failure times. In the first case the trend will tend to swamp any autocorrelation structure. In the second case, if the number of small X_i-values between large inter-failure times is a constant, m, then autocorrelations with lags $i(m+1)$ for $i=1, 2, 3, \ldots$ will tend to be large. However, if, as would be much more likely in practice, m varies, the behaviour of the r_k is less predictable. Indeed, in either of these cases the plot of the cumulative number of failures is likely to be more revealing with regard to the main features of the data than the sample autocorrelations.

Example 8.3 (continued)
Figure 8.4 shows the lag one plot for the Grampus data. This is rather typical of repairable systems data; see also Cox and Lewis (1966, p. 12). Figure 8.5

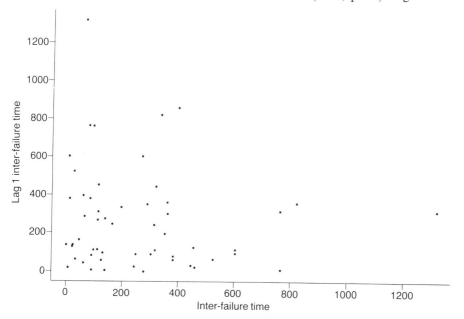

Figure 8.4 Lag 1 plot for the inter-failure times for the USS Grampus data.

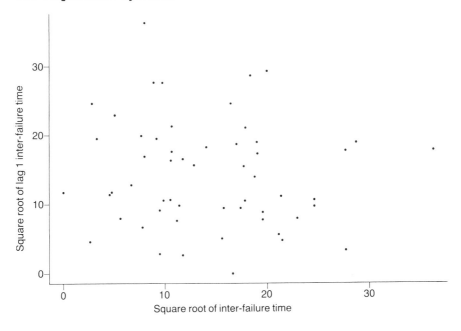

Figure 8.5 Lag 1 plot for the inter-failure times for the USS Grampus data after a square root transformation.

shows the same plot after a square root transformation. With the exception of a possible outlying point, which, of course, appears twice in the plot, there does not appear to be any marked lag-one autocorrelation structure.

The sample autocorrelations for lags one to ten are given in Table 8.4. These results indicate that there is little evidence in the data of marked autocorrelation structure. The corresponding sample autocorrelations for the data after square root transformation are very similar.

Table 8.4 Sample autocorrelations for the Grampus data of Example 8.3

k	1	2	3	4	5	6	7	8	9	10
r_k	-0.1	0.2	-0.1	-0.1	-0.1	-0.1	0.1	0.1	0.2	-0.1

8.5 NON-HOMOGENEOUS POISSON PROCESS MODELS

A stochastic point process may be regarded as a sequence of highly localized events distributed in a continuum according to some probabilistic mechanism. If we take the continuum to be time and the events as failures, it is clear that this is a natural framework for repairable systems. Cox and Lewis (1966) discuss point processes at length and Ascher and Feingold (1984), Thompson

(1988) and Ansell and Phillips (1989) consider applications specifically in reliability. The work of Triner and Phillips (1986), who analyse the data of Example 8.2 from a point processes viewpoint, is also of interest in this context, though see the discussion by Sweeting (1989).

Whilst it is possible to postulate a variety of point process models for the failures of a repairable system, we shall concentrate on the non-homogeneous Poisson process (NHPP). This model is conceptually simple, it can model happy and sad systems and the relevant statistical methodology is well-developed and easy to apply.

The assumptions for a NHPP are as for the (homogeneous) Poisson process discussed in Chapter 1 except that the ROCOF varies with time, rather than being constant. Consider a NHPP with time-dependent ROCOF $v(t)$ (this is sometimes called the **peril rate** or **intensity function**), then the number of failures in the time interval $(t_1, t_2]$ has a Poisson distribution with mean

$$\int_{t_1}^{t_2} v(t)\, dt.$$

Thus the probability of no failures in (t_1, t_2) is

$$\exp\left\{-\int_{t_1}^{t_2} v(t)\, dt\right\}.$$

Taking $v(t) \equiv v$ for all t just gives the homogeneous Poisson process with constant ROCOF (failure rate) v.

By choosing a suitable parametric form for $v(t)$, one obtains a flexible model for the failures of a repairable system in a 'minimal repair' set-up; that is, when only a small proportion of the constituent parts of the system are replaced on repair.

Suppose we observe a system for the time interval $(0, t_0]$ with failures occurring at $t_1, t_2, \ldots, t_n$. The likelihood may be obtained as follows. The probability of observing no failures in $(0, t_1)$, one failure in $(t_1, t_1 + \delta t_1)$, no failures in $(t_1 + \delta t_1, t_2)$, one failure in $(t_2, t_2 + \delta t_2)$ and so on up to no failures in $(t_n + \delta t_n, t_0)$ is for small $\delta t_1, \delta t_2, \ldots, \delta t_n$

$$\left\{\exp\left(-\int_0^{t_1} v(t)\, dt\right)\right\} v(t_1)\delta t_1 \left\{\exp\left(-\int_{t_1 + \delta t_1}^{t_2} v(t)\, dt\right)\right\} v(t_2)\delta t_2 \ldots$$

$$\left\{\exp\left(-\int_{t_n + \delta t_n}^{t_0} v(t)\, dt\right)\right\}.$$

Dividing through by $\delta t_1 \delta t_2 \ldots \delta t_n$ and letting $\delta t_i \to 0$ $(i = 1, 2, \ldots, n)$ gives the likelihood function.

$$L=\left\{\prod_{i=1}^{n} v(t_i)\right\} \exp\left\{-\int_{0}^{t_0} v(t)\,dt\right\}$$

(8.2)

and the log-likelihood is thus

$$l=\sum_{i=1}^{n} \log v(t_i)-\int_{0}^{t_0} v(t)\,dt$$

(8.3)

Another possible scheme for observation of a repairable system is to observe the system until the nth failure. Expressions (8.2) and (8.3) still apply, but with t_0 replaced by t_n. Sometimes, as in Example 8.2, the failure times are not observed and only the numbers of failures within non-overlapping time intervals are available. Suppose, for example, that $n_1, n_2, \ldots, n_m$ failures have been observed in non-overlapping time intervals $(a_1, b_1], (a_2, b_2], \ldots, (a_m, b_m]$. Then the likelihood is

$$L=\exp\left\{-\sum_{i=1}^{m} \int_{a_i}^{b_i} v(t)\,dt\right\} \prod_{i=1}^{m} \frac{\left(\int_{a_i}^{b_i} v(t)\,dt\right)^{n_i}}{n_i!}$$

Thus, the log-likelihood is, apart from an additive constant,

$$l=\sum_{i=1}^{m} \left\{n_i \log \int_{a_i}^{b_i} v(t)\,dt - \int_{a_i}^{b_i} v(t)\,dt\right\}$$

(8.4)

Therefore, once $v(t)$ has been specified, it is straightforward to use the likelihood-based methods of Chapter 3 to obtain ML estimates for any unknown parameters inherent in the specification of $v(t)$. We shall concentrate on two simple choices of $v(t)$ that give monotonic ROCOF:

$$v_1(t)=\exp(\beta_0+\beta_1 t)$$
$$v_2(t)=\gamma\delta t^{\delta-1}$$

where $\gamma>0$ and $\delta>0$.

Cox and Lewis (1966) discuss the NHPP with ROCOF $v_1(t)$ at length. It gives a simple model to describe a happy system ($\beta_1<0$) or a sad system ($\beta_1>0$). Also, if β_1 is near zero, $v_1(t)$ approximates a linear trend in ROCOF over short time periods. The NHPP with ROCOF v_1 will be discussed in more detail in section 8.6.

The NHPP with ROCOF $v_2(t)$ is referred to as the Weibull or power law process; see Crow (1974) and Ascher and Feingold (1984). The name Weibull process is a misnomer here since only the time to **first** failure has a Weibull distribution. Once again this model may represent a sad system $(\delta > 1)$ or a happy system $(0 < \delta < 1)$. The case $\delta = 2$ gives a linearly increasing ROCOF. The power law process will be discussed more fully in section 8.7.

Other, more complicated forms for $v(t)$ have been suggested in the literature. For example, a NHPP with $v(t)$ a hybrid of $v_1(t)$ and $v_2(t)$ has been suggested by Lee (1980), and generalizations of $v_1(t)$ to include polynomial terms in t are discussed in MacLean (1974) and Lewis and Shedler (1976).

8.6 NHPP WITH LOG-LINEAR ROCOF

In this section we give some likelihood-based statistical methods for fitting a NHPP with ROCOF $v_1(t)$ to a set of repairable systems data.

Putting $v_1(t) = \exp(\beta_0 + \beta_1 t)$ in (8.3), the log-likelihood for a repairable system observed for the time interval $(0, t_0)$ with failures at $t_1, t_2, \ldots, t_n$ is

$$l_1 = n\beta_0 + \beta_1 \sum_{i=1}^{n} t_i - \frac{\exp(\beta_0)\{\exp(\beta_1 t_0) - 1\}}{\beta_1}. \tag{8.5}$$

To obtain the ML estimates of β_0 and β_1 one can maximize (8.5) using, for example, a quasi-Newton method as supplied by NAG or Press et al. (1986). Alternatively, one can differentiate (8.5) with respect to β_0 and β_1 and then set the derivatives to zero. After some algebra $\hat{\beta}_0$ may be expressed explicitly in terms of $\hat{\beta}_1$. From here $\hat{\beta}_0$ may be eliminated from $\partial l_1 / \partial \beta_1 = 0$. Thus, (8.5) may be maximized by solving

$$\sum_{i=1}^{n} t_i + n\beta_1^{-1} - nt_0\{1 - \exp(-\beta_1 t_0)\}^{-1} = 0 \tag{8.6}$$

to give $\hat{\beta}_1$ and then taking

$$\hat{\beta}_0 = \log\left\{\frac{n\hat{\beta}_1}{\exp(\hat{\beta}_1 t_0) - 1}\right\}. \tag{8.7}$$

The programming involved in solving (8.6) is modest. A simple interactive search or repeated bisection will suffice. Possible starting values for the iterative scheme are discussed in section 8.8.

The observed information matrix, evaluated at the MLE, has entries

$$\frac{-\partial^2 l_1}{\partial \beta_0^2} = n.$$

$$\frac{-\partial^2 l_1}{\partial \beta_0 \partial \beta_1} = \sum_{i=1}^{n} t_i.$$

$$\frac{-\partial^2 l_1}{\partial \beta_1^2} = \hat{\beta}_1^{-1} \left\{ \sum_{i=1}^{n} t_i(\hat{\beta}_1 t_0 - 2) + nt_0 \right\}.$$

Inverting the information matrix gives the variance-covariance matrix for $(\hat{\beta}_0, \hat{\beta}_1)$. In particular the standard error of $\hat{\beta}_1$ is

$$\text{se}(\hat{\beta}_1) = \left\{ \hat{\beta}_1^{-1} \left\{ \sum_{i=1}^{n} t_i(\hat{\beta}_1 t_0 - 2) + nt_0 \right\} - n^{-1} \left(\sum_{i=1}^{n} t_i \right)^2 \right\}^{-1/2}. \tag{8.8}$$

The maximized log-likelihood may be obtained by substituting $\hat{\beta}_0$ and $\hat{\beta}_1$ in (8.5), giving

$$l_1(\hat{\beta}_0, \hat{\beta}_1) = n\hat{\beta}_0 + \hat{\beta}_1 \sum_{i=1}^{n} t_i - n. \tag{8.9}$$

A natural hypothesis to test in repairable systems reliability is that the ROCOF is constant; that is, $\beta_1 = 0$. If β_1 is set to zero, (8.5) becomes

$$n\beta_0 - t_0 \exp(\beta_0).$$

This is maximized when β_0 is $\log(n/t_0)$. So the maximized log-likelihood when $\beta_1 = 0$ is

$$n \log(n/t_0) - n.$$

Hence, using the results of section 3.4,

$$W = 2 \left\{ n\hat{\beta}_0 + \hat{\beta}_1 \sum_{i=1}^{n} t_i - n \log(n/t_0) \right\} \tag{8.10}$$

has an approximate $\chi^2(1)$ distribution when $\beta_1 = 0$. Large values of W supply evidence against the null hypothesis. Note that it is necessary to evaluate the MLEs in order to perform the test based on W.

A more commonly used test of $\beta_1 = 0$ is **Laplace's test**. This is based on the statistic

$$U = \frac{\sum_{i=1}^{n} t_i - \frac{1}{2} n t_0}{t_0 (n/12)^{1/2}}, \tag{8.11}$$

which has approximately a standard Normal distribution under the null hypothesis. If the alternative hypothesis is that $\beta_1 \neq 0$, then large values of $|U|$ supply evidence against the null hypothesis in favour of the general alternative. If the alternative is that $\beta_1 > 0$, a large value of U supplies evidence against the null hypothesis in favour of this alternative. If the alternative is that $\beta_1 < 0$, a large value of $-U$ supplies evidence against the null hypothesis in favour of the alternative. Laplace's test is asymptotically equivalent to the test based on W, but it does not require explicit evaluation of the MLEs. For further details see Cox and Lewis (1966, Chapter 3).

In the case in which the repairable system is observed until the nth failure, Laplace's test statistic U must be modified slightly. In (8.11) t_0 must be replaced by t_n, and n must be replaced by $n-1$.

When only numbers of failures within non-overlapping time intervals are available, then (8.4) becomes

$$\sum_{i=1}^{m} n_i \left\{ \beta_0 + \log\left(\frac{e^{\beta_1 b_i} - e^{\beta_1 a_i}}{\beta_1} \right) \right\} - e^{\beta_0} \sum_{i=1}^{m} \left(\frac{e^{\beta_1 b_i} - e^{\beta_1 a_i}}{\beta_1} \right). \tag{8.12}$$

In the special case in which the intervals are contiguous from time a_1 up to time b_m so that $b_1 = a_2, b_2 = a_3, \ldots, b_m = a_{m+1}$ the final term in (8.12) simplifies to

$$- e^{\beta_0} \left(\frac{e^{\beta_1 a_{m+1}} - e^{\beta_1 a_1}}{\beta_1} \right).$$

In this case the MLEs for β_0 and β_1 may be found first by solving

$$\sum_{i=1}^{m} n_i \left(\frac{a_{i+1} e^{\beta_1 a_{i+1}} - a_i e^{\beta_1 a_i}}{e^{\beta_1 a_{i+1}} - e^{\beta_1 a_i}} \right) - n \frac{(a_{m+1} e^{\beta_1 a_{m+1}} - a_1 e^{\beta_1 a_1})}{(e^{\beta_1 a_{m+1}} - e^{\beta_1 a_1})} = 0,$$

where $n = n_1 + n_2 + \cdots + n_m$, to give $\hat{\beta}_1$. Iterative methods as for solving (8.6) may be used here. Then

$$\hat{\beta}_0 = \log\left(\frac{n \hat{\beta}_1}{e^{\beta_1 a_{m+1}} - e^{\beta_1 a_1}} \right).$$

The NHPP with log-linear ROCOF also arises naturally as follows. Suppose a system has an unknown number of faults N. Each fault causes the system to fail, at which time the fault is located and removed. If the times at which the N failures occur are assumed to be independent exponential random variables with common mean, and if N is assumed to have a Poisson distribution, then the failure times arise according to a NHPP with ROCOF $v_1(t)$. Here the mean of the exponential distribution is the reciprocal of $-\beta_1$. This model has been much studied in the field of software reliability; see Jelinski and Moranda (1972) and Raftery (1988). Estimation of parameters in the software reliability context is just as outlined above, though note that β_1 must be negative. However, problems arise in a likelihood-based analysis if interest centres on estimating the number of remaining faults. In particular the estimated number of remaining faults can be infinite with substantial probability. In this context a Bayesian approach is more fruitful; see Raftery (1988). We shall not pursue this aspect further here. Software reliability has a vast specialist literature of its own. The interested reader is referred to Goel (1985), Langberg and Singpurwalla (1985), Littlewood and Verrall (1981) and Musa (1975).

Example 8.3 (continued)
Consider once again the USS Grampus data. Figure 8.3 is roughly linear, suggesting a roughly constant ROCOF. We can test the constant ROCOF hypothesis by embedding the model within a NHPP model with log-linear intensity. Suppose that the system was observed until just after the failure at 15070 hours. We shall fit the NHPP model with intensity $v_1(t)$ and then test whether $\beta_1 = 0$.

We first rescale the failure time by a factor 15070 so that on this new time scale $t_0 = 1$. This aids convergence of (8.6). On the new time scale we have

$$\sum_{i=1}^{56} t_i = 461731/15070 = 30.639.$$

Hence $\hat{\beta}_1 = 0.57$ and $\hat{\beta}_0 = 3.73$. Using (8.8), we obtain the standard error of $\hat{\beta}_1$ as 0.47. This suggests that $\hat{\beta}_1$ is not significantly different from zero. More formally (8.10) and (8.11) yield $W = 1.51$ and $U = 1.22$, neither of which is statistically significant at the 10% level. Thus, the constant ROCOF hypothesis cannot be rejected, confirming the impression given by Figure 8.3.

Example 8.4
Table 8.5 shows the intervals in days between successive failures of a piece of software developed as part of a large data system. These data have been

analysed by Jelinski and Moranda (1972) and Raftery (1988). In the software reliability context, two aspects are of interest. Firstly, is there any evidence of reliability growth? In other words is there evidence that $\beta_1 < 0$? Secondly, how many faults remain? We shall only tackle the first question here.

Table 8.5 Intervals in days between successive failures of a piece of software. The first interval is 9, the second interval is 12, and so on

9	12	11	4	7	2	5	8	5	7	1	6
1	9	4	1	3	3	6	1	11	33	7	91
2	1	87	47	12	9	135	258	16	35		

We suppose that observation of the software was made for 850 days. Therefore, $n = 34$, $t_0 = 850$ and

$$\sum_{i=1}^{34} t_i = 7015.$$

Hence, using (8.11) $U = -5.20$, which is highly significant, indicating that the hypothesis $\beta_1 = 0$ may be rejected in favour of $\beta_1 < 0$.

8.7 NHPP WITH ROCOF v_2

In this section we give some likelihood-based methods for fitting a NHPP with ROCOF $v_2(t)$ to a set of repairable systems data. Putting

$$v_2(t) = \gamma \delta t^{\delta - 1}$$

in (8.3) with $\gamma > 0$, $\delta > 0$ for a repairable system observed for the time period $(0, t_0]$ with failures at $t_1, t_2, \ldots, t_n$ gives log-likelihood

$$l_2 = n \log \gamma + n \log \delta + (\delta - 1) \sum_{i=1}^{n} \log t_i - \gamma t_0^\delta.$$

For this model closed-form MLEs for γ and δ are available:

$$\hat{\delta} = \frac{n}{n \log t_0 - \sum_{i=1}^{n} \log t_i}$$

and (8.13)

$$\hat{\gamma} = \frac{n}{t_0^{\hat{\delta}}}$$

The information matrix, evaluated at the MLE, has entries

$$\frac{-\partial^2 l_2}{\partial \gamma^2} = \frac{n}{\hat{\gamma}^2}$$

$$\frac{-\partial^2 l_2}{\partial \gamma \partial \delta} = t_0^{\hat{\delta}} \log t_0 \qquad (8.14)$$

$$\frac{-\partial^2 l_2}{\partial \delta^2} = \frac{n}{\hat{\delta}^2} + n (\log t_0)^2.$$

Note that if the time unit is chosen so that $t_0 = 1$, (8.13) simplifies to

$$\hat{\delta} = \frac{-n}{\sum\limits_{i=1}^{n} \log t_i}; \quad \hat{\gamma} = n.$$

In addition the information matrix with entries as in (8.14) becomes

$$\frac{-\partial^2 l_2}{\partial \gamma^2} = \frac{1}{n}$$

$$\frac{-\partial^2 l_2}{\partial \gamma \partial \delta} = 0$$

$$\frac{-\partial^2 l_2}{\partial \delta^2} = \frac{n}{\hat{\delta}^2}$$

Further, the standard error of $\hat{\delta}$ is

$$se(\hat{\delta}) = \frac{\hat{\delta}}{\sqrt{n}}.$$

We return now to the case of general t_0. The most commonly used test of constant ROCOF relative to the power law model is to test the null hypothesis that $\delta = 1$ using

$$V = 2 \sum_{i=1}^{n} \log\left(\frac{t_0}{t_i}\right). \qquad (8.15)$$

Under the null hypothesis, V has a $\chi^2(2n)$ distribution. Large values of V supply evidence against the null hypothesis in favour of reliability growth ($\delta < 1$). Small values of V are indicative of reliability deterioration ($\delta > 1$).

When the system is observed up to the nth failure, the formula (8.15) should be modified by replacing n by $n-1$, and t_0 by t_n. In addition the χ^2 degrees of freedom are $2(n-1)$ in place of $2n$. This gives the MIL-HDBK-189 (1981) test for a homogeneous Poisson Process against a NHPP alternative with monotonic intensity. As such it may be regarded as an alternative to Laplace's test.

An attractive feature of the NHPP with power law (and the reason for its misnomer of Weibull process) is that the time to first failure has a Weibull distribution. In particular

$$\Pr(T_1 > t) = \exp\left\{ -\int_0^t v_2(s)\, ds \right\}$$

$$= \exp(-\gamma t^\delta).$$

This is precisely the Weibull survivor function (2.3) with $\delta = \eta$ and $\gamma = \alpha^{-\eta}$. Note, however, that subsequent failure times or inter-failure times are not Weibull distributed. This neat link with a classical component reliability model and the tractability of the NHPP with power law intensity explain the popularity of this model in repairable systems reliability.

Example 8.5
Table 8.6 shows the times in operating hours to unscheduled maintenance actions for the number 3 main propulsion engine of the USS Halfbeak; see Ascher and Feingold (1984, p 75). Figure 8.6 shows the plot of i against t_i for these data. The plot suggests an increasing ROCOF.

We assume that the system was observed until the 71st failure at 25 518 hours. For these data $t_{71} = 25\,518$ and

Table 8.6 Failure times in operating hours for the number 3 main propulsion engine of USS Halfbeak

1 382	2 990	4 124	6 827	7 472	7 567	8 845	9 450	9 794
10 848	11 993	12 300	15 413	16 497	17 352	17 632	18 122	19 067
19 172	19 299	19 360	19 686	19 940	19 944	20 121	20 132	20 431
20 525	21 057	21 061	21 309	21 310	21 378	21 391	21 456	21 461
21 603	21 658	21 688	21 750	21 815	21 820	21 822	21 888	21 930
21 943	21 946	22 181	22 311	22 634	22 635	22 669	22 691	22 846
22 947	23 149	23 305	23 491	23 526	23 774	23 791	23 822	24 006
24 286	25 000	25 010	25 048	25 268	25 400	25 500	25 518	

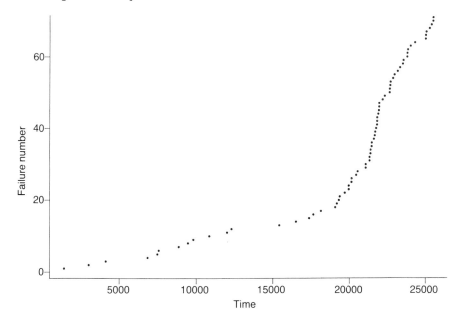

Figure 8.6 Plot of failure number against time in operating hours for the USS Halfbeak data.

$$\sum_{i=1}^{70} \log t_i = 684.58.$$

Hence $V = 51.44$. This is highly significant relative to $\chi^2(140)$ and confirms the impression given by Figure 8.6. Expression (8.13) gives $\hat{\delta} = 2.76$ with a standard error of 0.33.

8.8 CHOICE OF NHPP MODEL

In sections 8.6 and 8.7 we have given some likelihood-based methods for fitting NHPPs with ROCOFs $v_1(t)$ and $v_2(t)$. An obvious question is: how do we decide which of these models is preferable? More generally we are interested in choosing an appropriate form for $v(t)$. The approaches we shall discuss here are methods based on simple plots and likelihood-based methods.

Consider first some graphical methods. These are based either on the i versus t_i plot discussed briefly in section 8.4(a) or on $\hat{v}(t)$ given in (8.1). The former type of plot, which involves cumulative numbers of failures, is akin to working with $\hat{S}(t)$ or $\hat{H}(t)$ in the component reliability context or the ogive in classical statistics. Plots based on $\hat{v}(t)$ give localized information rather like the histogram in classical statistics.

Departures from linearity of a plot of i versus t_i are indicative of a non-constant ROCOF. Whether the ROCOF is increasing or decreasing may also be clear. For example, in Figure 8.6 it is obvious that the ROCOF is increasing. However, the precise form of $v(t)$ is not clear.

Now

$$E\{N(t)\} = \int_0^t v(y)\, \mathrm{d}y,$$

so when $v(t) = v_1(t)$,

$$E\{N(t)\} = \frac{e^{\beta_0}}{\beta_1}\{\exp(\beta_1 t) - 1\} \tag{8.16}$$

and when $v(t) = v_2(t)$,

$$E\{N(t)\} = \gamma t^\delta. \tag{8.17}$$

An obvious estimate of $E\{N(t)\}$ is the observed number of failures by time t. So, from (8.17) a plot of $\log i$ versus $\log t_i$ should be roughly linear with intercept $\log \gamma$ and slope δ if $v_2(t)$ is the appropriate ROCOF. An alternative plot of this type is the Duane plot; see Duane (1964). Here $\log(t_i/i)$ is plotted against $\log t_i$. Again, if $v_2(t)$ is appropriate the plot should be roughly linear but with slope $1 - \delta$. This plot was developed for situations in which reliability growth ($\delta < 1$) is expected, so that the expected slope is positive.

Unfortunately the form of (8.16) means that a similar approach cannot be used to check on the adequacy of the NHPP with ROCOF $v_1(t)$ without first fitting the model explicitly. The same is also true in general for more complicated functional forms of $v(t)$.

Now consider plots based on (8.1). If we divide the period of observation $(0, t_0]$ into k intervals $(0, a_1], (a_1, a_2], \ldots, (a_{k-1}, t_0]$, then an estimate of $v\{\frac{1}{2}(a_{j-1} + a_j)\}$ is

$$\hat{v}\{\tfrac{1}{2}(a_{j-1} + a_j)\} = \frac{N(a_j) - N(a_{j-1})}{a_j - a_{j-1}}, \tag{8.18a}$$

for $j = 1, 2, \ldots, k$ where $a_0 = 0$ and $a_k = t_0$. Thus a plot of $\hat{v}(b_j)$ versus b_j with $b_j = \frac{1}{2}(a_{j-1} + a_j)$ will give a reasonable indication of the shape of the ROCOF. The choice of k and the a_j is up to the user, just as in constructing a histogram. It is advisable to try several subdivisions in order to check that the visual impression given by the plot is not heavily dependent on the grouping used.

A plot of $\log \hat{v}(b_j)$ versus b_j should be roughly linear with slope β_1 and intercept β_0 if $v_1(t)$ is appropriate. This procedure may be used to supply starting estimates in obtaining the MLEs of β_0 and β_1; see (8.5) to (8.7). A plot of $\log \hat{v}(b_j)$ versus $\log b_j$ should be roughly linear with slope $\delta - 1$ and intercept $\log \gamma + \log \delta$ if $v_2(t)$ is appropriate.

In passing, it is worth noting another possible use for (8.18a). Suppose n failures have been observed in $(0, t_0]$. Then, on the assumption of a homogeneous Poisson process with constant ROCOF v, the estimated expected value of $N(a_j) - N(a_{j-1})$ is $n(a_j - a_{j-1})/t_0$. Thus, a classical χ^2 goodness-of-fit test with $k - 1$ degrees of freedom can be carried out for testing a homogeneous Poisson process against a general alternative. This procedure is less powerful than the tests based on U, V and W in sections 8.6 and 8.7 when the data have arisen from a NHPP with monotonic ROCOF, such as $v_1(t)$ or $v_2(t)$, but may be superior when the NHPP has non-monotonic ROCOF.

Example 8.3 (continued)
We have already seen that for the Grampus data we cannot reject the homogeneous Poisson process model in favour of the NHPP with ROCOF $v_1(t)$ or with ROCOF $v_2(t)$. Figure 8.7 shows a plot of $3014\hat{v}(b_j)$ versus b_j where $(0, 15\,070]$ has been subdivided into five equal intervals of length 3014 hours. The observed numbers of failures in these subintervals are $8, 13, 10, 7$ and 18. Under the homogeneous Poisson process assumption, the expected number of failures in each subinterval is 11.2. The χ^2 statistic, which is Σ (observed $-$ expected)^2/expected, is 7.04 on 4 degrees of freedom, which is not significant even at the 10% significance level. Similar results are obtained for other subdivisions of the time interval. These results tend to confirm those obtained earlier via (8.10) and (8.15).

As in section 3.4 comparison of maximized log-likelihoods of competing models may be worthwhile. In the case of nested NHPP models standard likelihood ratio methods may be used. However, when models are non-nested maximized log-likelihoods may still be useful for informal comparisons.

Suppose a NHPP with ROCOF $v(t; \theta)$ is thought to be a possible model for the data, where θ is an m-dimensional parameter to be estimated. Then the maximized log-likelihood is, from (8.3)

$$\hat{l} = \sum_{i=1}^{n} \log v(t_i; \hat{\theta}) - n,$$

where $\hat{\theta}$ is the MLE of θ. In particular, as in (8.9)

$$\hat{l}_1 = n\hat{\beta}_0 + \hat{\beta}_1 \sum_{i=1}^{n} t_i - n$$

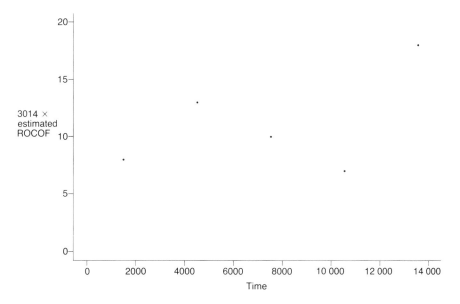

Figure 8.7 Plot of 3014 times the estimated ROCOF against time in operating hours for the USS Grampus data.

when $v(t) = v_1(t)$ and

$$\hat{l}_2 = n \log \hat{\gamma} + n \log \hat{\delta} + (\hat{\delta} - 1) \sum_{i=1}^{n} \log t_i - n \qquad (8.18b)$$

when $v(t) = v_2(t)$.

As mentioned in section 8.5, a number of generalizations of $v_1(t)$ and $v_2(t)$ have been considered in the literature. An attractive model which contains $v_1(t)$ and $v_2(t)$ as special cases is the NHPP with ROCOF

$$v_3(t) = \gamma \delta t^{\delta - 1} e^{\beta t}, \qquad (8.19)$$

with $\gamma > 0$, $\delta > 0$ and $-\infty < \beta < \infty$; see Lee (1980). Putting $\delta = 1$ gives $v_1(t)$ and putting $\beta = 0$ gives $v_2(t)$. The ROCOF (8.19) can be either monotonic (increasing or decreasing), or non-monotonic with one turning point. A particularly interesting situation occurs when $\delta < 1$ and $\beta > 0$ in which case $v_3(t)$ is bath-tub shaped. The NHPP with ROCOF $v_3(t)$ may be a useful model in its own right or simply as a means of providing a framework in which to choose between NHPPs with ROCOFs $v_1(t)$ and $v_2(t)$.

ROCOF $v_3(t)$ leads to a log-likelihood

$$l_3 = n(\log \gamma + \log \delta) + (\delta - 1) \sum_{i=1}^{n} \log t_i + \beta \sum_{i=1}^{n} t_i - \gamma \delta \int_0^{t_0} t^{\delta - 1} e^{\beta t} \, dt. \quad (8.20)$$

MLEs may be obtained from (8.20) by, for example, a quasi-Newton iterative scheme, the only complication being that the integral in (8.20) must in general be evaluated numerically. This can be achieved via standard quadrature methods (available in subroutine libraries such as NAG) or by using the fact that the integral is related to an incomplete gamma integral. For example, if $\beta < 0$

$$\int_0^{t_0} t^{\delta - 1} e^{\beta t} \, dt = (-\beta)^{-\delta} \int_0^{-\beta t_0} y^{\delta - 1} e^{-y} \, dy. \quad (8.21)$$

The integral on the right-hand side of (8.21) is an incomplete gamma integral; see Abramowitz and Stegun (1972) for details and Lau (1980) or Griffiths and Hill (1985, pp. 203–5) for a Fortran subroutine.

Lee (1980) discusses some tests for $\delta = 1$ in (8.19) and for $\beta = 0$ in (8.19). He discusses the null properties of the tests, but no details of the powers of the tests are given.

Example 8.5 (continued)
Earlier we fitted a NHPP model with power law ROCOF to the data for an engine of USS Halfbeak. In the light of the discussions above it is reasonable to ask whether the NHPP model with log-linear ROCOF fits the data any better. We first use some plots featured in this section to compare the relative merits of $v_1(t)$ and $v_2(t)$ for these data.

Table 8.7 shows $\hat{v}(t)$ and related quantities for the Halfbeak data. The

Table 8.7 Estimates of $v(t)$ and related quantities for the Halfbeak data

Time interval	b_j	Number of failures	$5000\,\hat{v}(b_j)$	$\log \hat{v}(b_j)$
(0– 5 000]	2 500	3	3	−7.42
(5 000–10 000]	7 500	6	6	−6.73
(10 000–15 000]	12 500	3	3	−7.42
(15 000–20 000]	17 500	12	12	−6.03
(20 000–25 000]	22 500	41	41	−4.80
(25 000–25 518]	25 259	6	57.92	−4.46

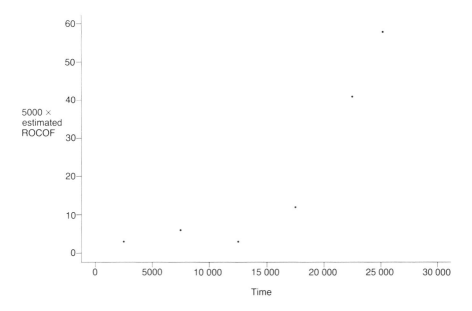

60
50
40
5000 ×
estimated
ROCOF   30
20
10
0

0   5000   10 000   15 000   20 000   25 000   30 000

Time

Figure 8.8 Plot of 5000 times the estimated ROCOF against time in operating hours for the USS Halfbeak data.

subdivision of the interval of observation $(0, 25\,518]$ has been made rather coarse deliberately to avoid having very low observed numbers of failures in the early sub-intervals, which upon being logged would have exaggerated visual impact.

Figure 8.8 shows a plot of $5000\hat{v}(b_j)$ versus b_j. As in Figure 8.6 it is clear that the overall trend in the ROCOF is upwards. Figure 8.9 shows a plot of $\log 5000\,\hat{v}(b_j)$ versus b_j. Figure 8.10 shows a plot of $\log 5000\,\hat{v}(b_j)$ versus $\log b_j$. Of these two plots, Figure 8.9 appears the more linear, which suggests that a log-linear ROCOF may be preferable to a power law ROCOF.

We now investigate the relative merits of the two models by obtaining the maximized log-likelihoods. We have already seen that by rescaling the time unit so that $t_0 = 1$, we obtain MLEs $\hat{\gamma} = 71$ and $\hat{\delta} = 2.76$ for the NHPP with power law ROCOF. This gives maximized log-likelihood (8.18b), $\hat{l}_2 = 258.47$.

Using the rescaled time unit and solving (8.5) yields MLEs $\hat{\beta}_0 = 1.81$ and $\hat{\beta}_1 = 3.81$ for the NHPP with log-linear ROCOF. Using (8.8), we obtain a standard error of 0.56 for $\hat{\beta}_1$. For this model the maximized log-likelihood, (8.9), is $\hat{l}_1 = 263.16$, which is considerably greater than $\hat{l}_2$. This confirms the impression given by Figures 8.9 and 8.10 that the NHPP with ROCOF $v_1(t)$ is preferable here.

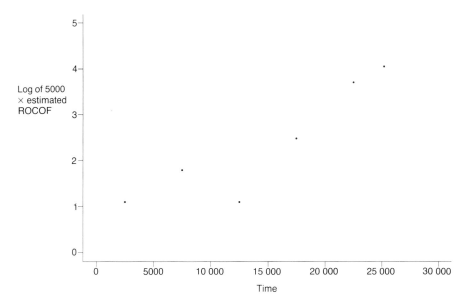

Figure 8.9 Plot of the logarithm of 5000 times the estimated ROCOF against time in operating hours for the USS Halfbeak data.

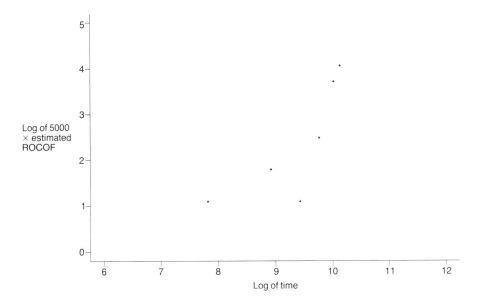

Figure 8.10 Plot of the logarithm of 5000 times the estimated ROCOF against the logarithm of time in operating hours for the USS Halfbeak data.

8.9 DISCUSSION

In the previous sections of this chapter the emphasis has been on very simple statistical methods. In many circumstances we believe that graphical methods or the fitting of NHPP models with smoothly changing ROCOF will be adequate. However, there are some situations which lead to somewhat more difficult statistical problems that cannot be covered by these simple methods.

For example, there may be a sudden change in ROCOF rather than a smooth change. If the time at which this change occurs is unknown, we have the **change-point** problem. This might arise when an electronic system suffers a sudden surge in electrical supply which causes the reliability of the system to be impaired thereafter. Typically, we are interested in estimating the time at which the change took place. Useful references are Akman and Raftery (1986) and Raftery and Akman (1986).

Another area which we have not covered is that in which covariate information is available. This approach is dealt with briefly in Cox and Lewis (1966, Chapter 3). A recent and important reference in this topic is Lawless (1987), where he adopts an approach akin to proportional hazards modelling in both parametric and semi-parametric frameworks.

For further discussion of topics of current interest in repairable systems reliability the reader is referred to the paper by Ansell and Phillips (1989) and the resulting discussion, and to Ascher and Feingold (1984).

9

Models for system reliability

9.1 INTRODUCTION

For most of this book we have discussed the reliability of individual units without reference to their place within the overall structure of the system under study. For most large systems this is of course much too simple. Instead, the performance of the system can be analyzed as a function of individual components. If data are available on the individual components, then it is possible to make statistical inferences about the reliability of those components, but this still leaves the problem of calculating system reliability from the individual component reliabilities. The purpose of this chapter is to review some of the models that may be adopted for this purpose.

A detailed study of the probabilistic relations between components and systems inevitably involves a higher level of mathematical sophistication than that assumed for most of this book. The main emphasis of this chapter is to describe the models, and to give an outline of the mathematical derivation, with references for the reader desiring the full mathematical details. Nevertheless, there is more mathematics in this chapter than elsewhere in the book. The reader desiring a quick overview is advised to concentrate on sections 9.1–4, skimming the remaining sections to appreciate the range of models being studied but not necessarily trying to understand the mathematical development for each one.

The best known class of reliability models is that of **coherent systems**. The fundamental concept of a coherent system is that the individual components are in one of two states, functioning or failed, and the state of the system is represented in terms of the states of individual components through what is called the **structure function**. Two key properties are (a) relevance of every component, i.e. there is no component whose reliability does not affect the reliability of the system, and (b) monotonicity; loosely, the concept that the reliability of the system can never be improved by one of its components becoming less reliable. Examples include series and parallel systems, **k out of n** systems and a whole host of network reliability problems. The topic of coherent systems is presented in section 9.2.

The reader wanting a more detailed review is advised to consult the book by Barlow and Proschan (1981).

Some interesting statistical problems arise in connection with coherent systems. Point estimation is generally no problem, because the reliability of the system may be estimated by substituting the point estimates of reliabilities of individual components. However, there are much more subtle problems connected with interval estimates, both classical and Bayesian. This topic is reviewed in section 9.3.

One objection to the concept of a coherent system is that it only allows two states, failed or unfailed, for each component. A generalisation is **multi-state systems**, in which each component is classified into one of M states with specified probabilities. This topic is reviewed in section 9.4.

A different approach to system reliability is via the concept of **load sharing**. The basic idea is that there is some fixed load on the system; in strength and fatigue problems this is usually a mechanical load, but the concept is also applicable in other situations such as electrical networks. The components, however, have random strengths or random failure times (or, in the electrical case, random failure currents) and it therefore follows that some of them will fail under the applied load. This load is then redistributed onto the other components according to a load-sharing rule. The load-sharing rule is crucial to the calculation of system reliability; the reliability of a load-sharing system depends on the load-sharing rule in much the same way as the reliability of a coherent system depends on the structure function. Two particular classes of load-sharing rule have been studied in detail. The first is **equal load-sharing**, which as the name implies means that load is shared equally over all components, though there are some situations, such as fibres with random slack, which use the same principle without actually leading to equality of load on all components. The second class of load-sharing rules is **local load-sharing**, in which the load on a failed component is distributed to nearby unfailed neighbours. These models are reviewed in sections 9.5–11, and in the final section 9.12 we briefly review some statistical applications of them.

9.2 COHERENT SYSTEMS

Consider a system with n components. Let x_i denote the state of component i: x_i is 1 if the component is functioning, 0 if it is failed. Similarly, let ϕ denote the state of the system, 1 or 0. We consider cases in which the state of the system is determined by the state of the individual components. That is,

$$\phi = \phi(x)$$

where $x = (x_1, \ldots, x_n)$ is the vector of component states.

Example 9.1
A **series system** is one in which all components must be functioning for the system to function. In other words, ϕ is 1 if and only if all the x_i are 1. So

$$\phi(x) = \prod_{i=1}^{n} x_i.$$

Example 9.2
A **parallel system** is one in which only one component needs to be functioning in order for the system to function. In this case

$$\phi(x) = 1 - \prod_{i=1}^{n} (1 - x_i).$$

Example 9.3
One generalization which encompasses both series and parallel systems is a **k out of n** system in which the system is functioning if at least k out of the n components are functioning. In this case, $\phi(x)$ is 1 if $\sum x_i \geq k$, 0 if $\sum x_i < k$. Obviously, a series system corresponds to the case $k = n$, and a parallel system to $k = 1$.

Example 9.4
This is an example of a **network reliability** problem in which the system may be represented by a network of components, and the state of the system by the existence or otherwise of a path through the functioning components.

A computer system consists of one central computer to which there are three terminals. The computer has a printer attached to it, but it is also possible to obtain printed output by spooling a file to another, centrally located and maintained, printer. The system is considered to be functioning if it is possible to use the computer up to the point of obtaining printed output. For this we require that (a) the computer itself is functioning, (b) at least one of the three terminals is functioning and available, and (c) either the computer's own printer or the link to the central printer is available. The system may be represented pictorially by Figure 9.1 in which components 1, 2 and 3 are the three terminals, 4 the computer itself, 5 the local printer and 6 the link to the central computer. In this case

$$\phi(x) = \{1 - (1 - x_1)(1 - x_2)(1 - x_3)\} x_4 \{1 - (1 - x_5)(1 - x_6)\}. \tag{9.1}$$

Although this is still quite a simple example, one can see here the potential for much bigger things. For instance, a whole computer system for a company or a university could in principle be represented by a network diagram of this

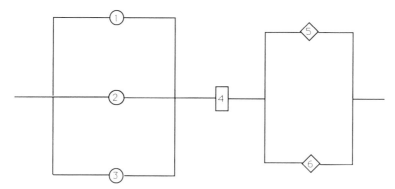

Figure 9.1 Diagram of computer system (example 9.4).

form, though a large system might require thousands of components and a very complicated network structure. Even nuclear power stations have been modelled by networks of this form.

The function ϕ is known as the **structure function**. The system represented by a structure function is **coherent** if it has the two additional properties mentioned in the introduction: relevance of every component, and monotonicity. The ith component is **irrelevant** if, for all states $x_1, \ldots, x_{i-1}, x_{i+1}, \ldots, x_n$ of the other components, the state of the system is the same whether x_i is 0 or 1:

$$\phi(x_1, \ldots, x_{i-1}, 0, x_{i+1}, \ldots, x_n) = \phi(x_1, \ldots, x_{i-1}, 1, x_{i+1}, \ldots, x_n).$$

A component which is not irrelevant is **relevant**. **Monotonicity** of a structure function simply refers to its monotonicity in every x_i, or equivalently

$$\phi(x_1, \ldots, x_{i-1}, 0, x_{i+1}, \ldots, x_n) \le \phi(x_1, \ldots, x_{i-1}, 1, x_{i+1}, \ldots, x_n). \tag{9.2}$$

Definition
A structure function ϕ defines a *coherent system* if it is monotone and every component is relevant.

One consequence of this definition is that, for each i, there must exist a set of $x_1, \ldots, x_{i-1}, x_{i+1}, \ldots, x_n$ for which the inequality in (9.2) is strict.

Probability calculations for coherent systems
We now turn to the question of computing failure probabilities for coherent systems. Our attention here is confined to general principles, and not to the efficient computational solution of these problems in large systems. For the latter the reader is referred to the book by Barlow and Proschan (1981) or, for reviews of more recent developments, the article by Agrawal and Barlow

(1984), Agrawal and Satyanarayna (1984) and Provan and Ball (1984), all of which appeared in a special issue of *Operations Research*.

Suppose, first, that the components are independent and p_i is the probability that the ith component is functioning. In principle, it is easy to write down an algebraic expression for the probability, $h(p)$, that the system is functioning in terms of $p = (p_1, \ldots, p_n)$:

$$h(p) = \sum_x \left[\phi(x) \prod_{i=1}^{n} \{ p_i^{x_i} (1 - p_i)^{1 - x_i} \} \right].$$

Here h is called the **reliability function**.

For example, with a series system we easily see that

$$h(p) = \prod_{i=1}^{n} p_i$$

and for a parallel system

$$h(p) = 1 - \prod_{i=1}^{n} (1 - p_i).$$

A k out of n system is not so easy to handle in general, but if $p_1 = \cdots = p_n = p$ it reduces to the Binomial distribution:

$$h(p) = \sum_{j=k}^{n} \binom{n}{j} p^j (1 - p)^{n-j}.$$

With Example 9.4, the reliability function is in effect the expectation of equation (9.2):

$$h(p) = \{ 1 - (1 - p_1)(1 - p_2)(1 - p_3) \} p_4 \{ 1 - (1 - p_5)(1 - p_6) \}. \tag{9.3}$$

In general, however, passing from the structure function to the reliability function is not as simple as that from (9.1) to (9.3). A useful concept (Barlow and Proschan) is that of **minimal path** and **minimal cut**. A path is any subset of components with the property that, if those components are functioning, then the entire system is functioning; likewise a *cut* is a set of components whose failure guarantees failure of the system. A path (cut) is minimal if no strict subset has the property of being a path (cut). Suppose a system has p minimal paths $P_1, \ldots, P_p$ and k minimal cuts $K_1, \ldots, K_k$. A general representation for the structure function is

$$\phi(x) = \max_{1 \le j \le p} \min_{i \in P_j} x_i = \min_{1 \le j \le k} \max_{i \in K_j} x_i. \tag{9.4}$$

In words, equation (9.4) says that a system functions if all the components on one of its paths function, or equivalently if at least one component in every cut functions. For Example 9.4, the minimal paths are $\{1, 4, 5\}$, $\{2, 4, 5\}$, $\{3, 4, 5\}$, $\{1, 4, 6\}$, $\{2, 4, 6\}$ and $\{3, 4, 6\}$. The minimal cuts are $\{1, 2, 3\}$, $\{4\}$ and $\{5, 6\}$. The two representations corresponding to (9.4) are

$$\phi(x) = \max\{\min(x_1, x_4, x_5), \min(x_2, x_4, x_5), \ldots, \min(x_3, x_4, x_6)\} \qquad (9.5)$$

$$\phi(x) = \min\{\max(x_1, x_2, x_3), x_4, \max(x_5, x_6)\}. \qquad (9.6)$$

It is obvious that (9.6) is the same as (9.1), but perhaps not so immediately obvious that (9.5) is.

For a general coherent system, either of the two representations in (9.4) may be taken to define the reliability function: if $X_1, \ldots, X_p$ represent independent random variables with $\Pr\{X_i = 1\} = 1 - \Pr\{X_i = 0\} = p_i$ then

$$h(p) = E\{\max_{1 \le j \le p} \min_{i \in P_j} X_i\} = E\{\min_{1 \le j \le k} \max_{i \in K_j} X_i\}. \qquad (9.7)$$

In simple cases such as (9.6) this leads directly to a calculation of the system reliability function. The reason that this particular example is simple is that the cuts are disjoint subsets of the components, so that the minimum over cuts is a minimum over independent random variables. Since

$$E\{\max_{i \in K_j} X_i\} = 1 - \prod_{i \in K_j} (1 - p_i)$$

this leads directly to

$$h(p) = \prod_{1 \le j \le k} \left\{1 - \prod_{i \in K_j} (1 - p_i)\right\}$$

which is equivalent to (9.3).

In the general case, (9.7) leads to a representation of the system state as either the maximum or the minimum of a set of dependent random variables. The **inclusion-exclusion formula** (Feller (1968), pp 98–101 or Barlow and Proschan (1981), pp 25–6) may be used to calculate the reliability function for the system, or equivalently the first few terms of that formula may be used to obtain inequalities. This formula is

$$\Pr\left\{\bigcup_{i \in N} A_i\right\} = \sum_{I \subseteq N, I \neq \emptyset} (-1)^{|I|+1} \Pr\left\{\bigcap_{i \in I} A_i\right\} \qquad (9.8)$$

for any set of events $\{A_i, i \in N\}$. In principle this is a general procedure for calculating the reliability of a coherent system, but of course it quickly becomes computationally intractable if the system is large.

Dependent components

The discussion so far has assumed independent components. In many contexts this is an unrealistic assumption. Computation of system reliability under assumptions of dependent components has been considered by Hagstrom and Mak (1987); they emphasize the increase of computational complexity caused by the dependence. Alternatively, it may be possible to obtain inequalities under assumptions of positive dependence similar to those mentioned in Chapter 7. The strongest property of this nature is **association**: a vector of random variables T is associated if

$$\text{cov}\{\Gamma(T), \Delta(T)\} \geq 0$$

for any pair of increasing binary functions Γ and Δ.

It is easy to see that, for associated random variables $T_1, \ldots, T_m$,

$$\Pr\{T_1 > t_1, \ldots, T_m > t_m\} \geq \prod_{j=1}^{m} \Pr\{T_j > t_j\},$$

$$\Pr\{T_1 \leq t_1, \ldots, T_m \leq t_m\} \geq \prod_{j=1}^{m} \Pr\{T_j \leq t_j\},$$

and hence it is possible to deduce

$$\prod_{j=1}^{k} \left\{1 - \prod_{i \in K_j} (1 - p_i)\right\} \leq h(p) \leq 1 - \left\{\prod_{j=1}^{p} \left(1 - \prod_{i \in P_j} p_i\right)\right\} \tag{9.9}$$

This inequality, of course, also holds in the case of independent components.

The idea of association and its consequences for coherent system reliability are discussed in much greater detail in Chapter 2 of Barlow and Proschan (1981).

9.3 ESTIMATION OF RELIABILITY FOR COHERENT SYSTEMS

The topic of interval estimation of system reliability is one with an extensive, but scattered, literature. A method of calculating exact confidence limits was formulated by Buehler (1957), and much of the subsequent literature has developed from that. Tables for this were compiled by Lipow and Riley (1960), but it has been generally assumed that exact computation is beyond the reach of routine users. Consequently, a wide literature on approximate solutions has grown up. From the perspective of contemporary computing, exact calculation of Buehler's confidence limits is less of a difficulty, at least for small problems of the kind considered in the literature. Therefore, it is perhaps time to consider whether Buehler's method is in fact the right way to formulate the problem. In particular, it is a problem which throws into rather sharp focus the difference between the classical and Bayesian ways of thinking.

In common with most of the literature, we restrict ourselves to the case of binomial test data, i.e. the data consisting of components classified as survived and failed. Consideration has also been given to data consisting of failure times, assuming an exponential failure time with unknown parameter for each type of component. See for example Mann and Grubbs (1974) and references therein. Lindley and Singpurwalla (1986a) also considered this problem but in a somewhat different context, namely how to pool different experts' opinions in a Bayesian analysis. There does not seem to have been much consideration of general failure distributions.

Suppose, then, there are r distinct types of component with survival probabilities $p_1, \ldots, p_r$. The reliability function for a coherent system may be written as a function $h(p_1, \ldots, p_r)$. The estimation problem for coherent systems is the problem of estimating h from information on the individual $p_1, \ldots, p_r$.

If $\hat{p}_1, \ldots, \hat{p}_r$ are point estimates of $p_1, \ldots, p_r$ then an obvious estimator of h is

$$\hat{h} = h(\hat{p}_1, \ldots, \hat{p}_r). \tag{9.10}$$

If the method of estimation is maximum likelihood, then (9.10) gives exactly the maximum likelihood estimate of h. However, the problem of interval estimation is more complicated. The delta method used in Chapters 3 and 4 (appendix) may be used to obtain an approximate standard error of $\hat{h}$ and hence a confidence interval via the Normal approximation, but that method often gives rise to rather a poor approximation, so our purpose here is to consider other possible constructions.

We examine first the possibilities for exact confidence limits. We consider only lower confidence limits, since in the reliability context this is usually what is of most interest, but of course the same reasoning may be used to derive upper confidence limits.

Consider first a single component which functions with probability p. Suppose n test specimens are observed of which Y fail. Assuming independence, Y has a Binomial distribution with parameters n and $1-p$. A test of the null hypothesis $p \leq p_0$ against the alternative $p > p_0$ has the following critical region: reject the null hypothesis if $Y \leq y^*$, where y^* is the largest value such that

$$\sum_{y=0}^{y^*} \binom{n}{y}(1-p_0)^y p_0^{n-y} \leq \alpha,$$

α being the prescribed type I error probability.

A $100(1-\alpha)\%$ lower confidence bound for p may be defined as the smallest value of p_0 which the above test of $p \leq p_0$ would accept. This is given by p^*, where p^* solves

$$\sum_{y=0}^{Y} \binom{n}{y}(1-p^*)^y(p^*)^{n-y} = \alpha. \tag{9.11}$$

Under repeated sampling for fixed n and p, p^* is a random variable satisfying

$$\Pr\{p^* \leq p\} \geq 1-\alpha. \tag{9.12}$$

The discrete nature of the random variable Y is responsible for the fact that (9.12) is only an inequality and not an equality.

This construction, exploiting the close relationship between tests and confidence intervals, relies on the fact that there is a natural 'tail' of values of y: any y less than the observed Y is 'more extreme' than Y, and hence is included in the rejection region of the test. This form of rejection region is implicit in (9.11).

We would like to extend this method to the estimation of system reliability. For most of this section, we confine ourselves to a simple series system since this illustrates in simple form the difficulties and possible solutions that may be considered for coherent systems in general. There is a fundamental difficulty in extending the idea just outlined from component to system reliability. Suppose, for example, we have a series system with r different types of components with reliabilities $p_1, \ldots, p_r$. Assume that n_i independent tests are carried out on components of type i and that Y_i of those test components are failed. The natural rejection region for a hypothesis of the form $p_1 \leq p_{1,0}, \ldots, p_r \leq p_{r,0}$ would consist of all vectors $(y_1, \ldots, y_r)$ which are 'more extreme' than the observed $(Y_1, \ldots, Y_r)$. However, this implies some ordering on the space of outcomes. For example, in the case $r=2$ we must decide whether $Y_1=1, Y_2=4$ is a 'worse' outcome than $Y_1=Y_2=3$. In general, the only way round this problem is to define arbitrarily an ordering function $g(y_1, \ldots, y_r)$ on the space of possible outcomes, with the interpretation that the lower g, the better the system. Let $(Y_1, \ldots, Y_r)$ denote the observed data vector.

Given such an ordering, Buehler's lower confidence limit is the minimum value of $h^* = p_1^* \ldots p_r^*$, such that

$$\sum_{y:g(y_1,\ldots,y_p) \leq g(Y_1,\ldots,Y_p)} \left[\prod_{i=1}^{r} \binom{n_i}{y_i} (1 - p_i^*)^{y_i} (p_i^*)^{n_i - y_i} \right] = \alpha. \tag{9.13}$$

In fact a fairly direct extension of the argument that leads to (9.11), shows that (9.13) gives a valid lower confidence limit in the case of system reliability. It remains to specify a suitable ordering on the space of observed vectors, but if the object of the exercise is to estimate system reliability h, then a natural ordering function is one minus estimated system reliability, i.e. $g(y_1, \ldots, y_p) = 1 - \hat{h} = 1 - \prod \{(n_i - y_i)/n_i\}$.

This method has been developed fairly extensively; Lipow and Riley (1960) gave tables for equal sample sizes and $r \leq 3$ and a number of authors have shown how to simplify the computations in specific cases; for a review see Winterbottom (1984). We have programmed up the minimization (9.13) using the NAG routine E04VDF for nonlinear optimization with nonlinear constraints. Also, it is clear that other kinds of system can be handled by using the correct structure function $h^*(p_1^*, \ldots, p_r^*)$ in (9.13). Exact computation appears to be feasible for $r \leq 3$ and n_i not too large, but quickly becomes prohibitive for large numbers of components or large sample sizes.

As an alternative to exact calculation, a number of approximate methods have been suggested:

1. Since the individual $\hat{p}_i$ have variances $\hat{p}_i(1 - \hat{p}_i)/n_i$, the delta method gives the approximate variance of $\hat{h} = \prod \hat{p}_i$ as $\hat{h}^2 \sum \{(1 - \hat{p}_i)/(n_i \hat{p}_i)\}$. This could be used directly, via a Normal approximation, to construct an asymptotic confidence interval. Easterling (1972) proposed a refinement which amounts to the following: replace the actual experiment by a hypothetical single binomial experiment with sample size $\hat{n}$ and observed number of failures $\hat{Y}$, so that both the estimated $\hat{h} = (\hat{n} - \hat{Y})/\hat{n}$ and its estimated variance $\hat{h}(1 - \hat{h})/\hat{n}$ equate to those obtained for the true experiment. It is not hard to see that this implies $\hat{n} = (1 - \hat{h})/[\hat{h} \sum \{Y_i/(n_i(n_i - Y_i))\}]$. The confidence region is then calculated as for a single binomial experiment. In general, $\hat{n}$ and $\hat{Y}$ will not be integers but they are then rounded up to the nearest integer. The method is obviously extendable to other reliability functions, if we substitute the appropriate estimate and standard error for system reliability.

2. Another approximate method, applicable when the p_is are near 1, is to use a Poisson approximation. In this case $1 - \hat{h} \approx \sum (1 - \hat{p}_i)$ and the individual $1 - \hat{p}_i = Y_i/n_i$, where Y_i has approximately a Poisson distribution. When the n_i are all equal this implies that $n_1(1 - \hat{h})$ has an approximate Poisson distribution with mean $n_1(1 - h)$ and this fact may be used directly to obtain confidence limits.

3. The likelihood ratio technique discussed in Chapters 3 and 4 is also applicable to system reliability problems. The likelihood ratio for system reliability h is given by

$$\Lambda(h) = \sup\{L(p_1, \ldots, p_r): \textstyle\prod p_i = h\}/\sup\{L(p_1, \ldots, p_r): \textstyle\prod p_i = \hat{h}\}$$

where L is the likelihood function and $\hat{h}$ the maximum likelihood estimate. An approximate $100(1-\alpha)\%$ confidence interval for h may be defined as $\{h: -2 \log \Lambda(h) \leq \chi^2_{1,\alpha}\}$, where $X^2_{1,\alpha}$ is the upper-α point of the χ^2_1 distribution. One-sided confidence limits may be obtained by a suitable one-sided modification of this. The method is more accurate than the simple form of delta method, but computation may be hard. However, it is of very general applicability. Madansky (1965) was an early proponent of this technique.

4. Mann and Grubbs (1974) (and earlier papers reviewed therein) proposed some analytic approximations essentially improving on the simple Normal approximation. Their results were accompanied by quite detailed numerical study, so this paper is worth studying for its comparisons of several different approaches. Yet another approximate method is the use of Edgeworth and Cornish–Fisher expansions to improve on the Normal asymptotic distribution of the maximum likelihood estimate. See Winterbottom (1980, 1984).

5. A halfway house between exact and approximate methods is the bootstrap. This takes a simple form: given n_i and Y_i generate a bootstrap $Y_{i,b}$ by simulation from a Binomial distribution with parameters n_i and Y_i/n_i. Repeat for all i, and hence form a bootstrap estimate of system reliability, $\hat{h}_b = \prod\{(n_i - Y_{i,b})/n_i\}$. Repeat for $b = 1, \ldots, B$ to form a bootstrap sample of size B, and then use the empirical distribution of $\hat{h}_b, b = 1, \ldots, B$ to set confidence limits. Although the modern form of the bootstrap is due to Efron (1979), the idea of using simulation for computing confidence intervals for system reliability goes back at least as far as Rosenblatt (1963). Other variants on the idea of running Monte Carlo experiments include the papers of Rice and Moore (1983) and Chao and Huwang (1987), each of whom proposed simulations with the component reliabilities varying from one replicate to the next to represent our uncertainty over their true values. Martz and Duran (1985) made a simulation comparison of confidence limits obtained by the bootstrap method with those obtained from a variant of the Lindstrom–Madden method (to be described in a moment), and with a Bayesian method. A surprising conclusion of Martz and Duran was that, under certain circumstances, the bootstrap can be highly anti-conservative.

Although the detailed discussion here has been confined to series systems, it is clear that many of the concepts go over to more general coherent structures.

The Lindstrom–Madden method
Among the various approximate schemes that have been developed, one that has attracted particular attention is that due to D. L. Lindstrom and J. H. Madden, which appears to have been first described in the book by Lloyd and Lipow (1962). A brief and readable account is due to Soms (1985), whom we follow here.

Suppose we have n_i observations on component i, ordered so that $n_1 \leq n_2 \leq \cdots \leq n_r$. Consider building a system by randomly selecting a single member from each type of component. In this way we can sample n_1 systems and the expected number of failures is $z_1 = n_1(1 - \hat{h})$. The idea is to treat this as a single component for which n_1 observations are taken resulting in z_1 failures. If z_1 is an integer then we just substitute z_1 for Y and n_1 for n in (9.11). To handle the case when z_1 is not an integer, let us first note the identity

$$\sum_{y=0}^{Y} \binom{n}{y}(1-p)^y p^{n-y} = I_p(n-Y, Y+1)$$

where I_p is the incomplete beta function

$$I_p(r, s) = \{B(r, s)\}^{-1} \int_0^p t^{r-1}(1-t)^{s-1}\, dt$$

and $B(r, s) = \Gamma(r)\Gamma(s)/\Gamma(r+s)$. Thus the method reduces to solving the equation

$$I_h(n_1 - z_1, z_1 + 1) = \alpha$$

the solution h defining the required lower confidence limit.

This appears originally to have been an *ad hoc* suggestion to approximate the solution of (9.13), but there has been considerable work done to justify the idea. Sudakov (1974) obtained bounds on the solution of (9.13) and showed that it always lies between the two solutions to the Lindstrom–Madden procedure obtained using z_1 and the integer part of z_1; in particular, if z_1 is an integer the method is exact. (Soms (1985) claimed that an error had been found in Sudakov's proof but that numerical evidence indicated the inequality still to be fulfilled in practice.)

Soms (1988) proposed an extension of the Lindstrom–Madden method to reliability functions of the form

$$h(\boldsymbol{p}) = \prod_{i=1}^{r} p_i^{\gamma_i}$$

for integers $\gamma_1, \ldots, \gamma_r$. In this case the evidence for the performance of the method appears to be entirely numerical. It is presumably an open question how far the method can be pushed in terms of handling more complicated reliability functions.

Overall, the Lindstrom–Madden method has been the subject of a lot of study, but even with Soms' (1988) extension it is still fairly restricted in scope and there is no evidence that other methods of approximation are necessarily inferior. The study by Martz and Duran (1985) suggested that the method can be much too conservative.

Bayesian methods.
Bayesian methods often come into their own when dealing with complicated likelihoods, and this is certainly true for problems of system reliability. Again we shall confine our detailed discussion to series systems though the concepts are applicable to coherent systems in general.

Suppose we again have binomial data on r types of component, the ith component with reliability p_i having n_i observations with Y_i observed failures. Suppose the prior distribution of p_i is a beta distribution with parameters α_i and β_i, i.e. the density of p_i is

$$f(p_i) = B(\alpha_i, \beta_i)^{-1} p_i^{\alpha_i - 1}(1 - p_i)^{\beta_i - 1}.$$

The assumption of a conjugate prior is made here purely for reasons of simplicity: it is not essential for the concepts. After the sample has been collected, the posterior distribution of p_i is again beta with parameters α_i^* and β_i^*, where

$$\alpha_i^* = \alpha_i + n_i - Y_i, \ \beta_i^* = \beta_i + Y_i$$

and these are independent for different i.

Point estimation of $h(\boldsymbol{p})$ is straightforward: the posterior mean is

$$E\{h(\boldsymbol{p})|\boldsymbol{Y}\} = \prod_i E\{p_i|Y_i\} = \prod_i \{\alpha_i^*/(\alpha_i^* + \beta_i^*)\}.$$

Similarly, the exact posterior variance is available from

$$\mathrm{Var}\{h(\boldsymbol{p})|\boldsymbol{Y}\} = \prod_i E\{p_i^2|\boldsymbol{Y}\} - \prod_i [E\{p_i|\boldsymbol{Y}\}]^2,$$

and

$$E\{p_i^2|Y\}=(\alpha_i^*+1)\alpha_i^*/\{(\alpha_i^*+\beta_i^*+1)(\alpha_i^*+\beta_i^*)\}.$$

Formulae can be derived for other reliability functions. It is also possible to calculate the Mellin transform of the posterior distribution and hence to obtain the complete posterior distribution by numerical application of the inversion formula for Mellin transforms; this was first suggested by Springer and Thompson (1966).

The posterior probability that $h \le h_0$ is given by

$$\int_0^1 \cdots \int_0^1 I\left(\prod_i p_i \le h_0\right)\prod_i \{B(\alpha_i^*, \beta_i^*)^{-1} p^{\alpha_i^*-1}(1-p_i)^{\beta_i^*-1}\}\, dp_1 \ldots dp_r, \quad (9.14)$$

I denoting the indicator function.

This can be simplified as follows: the posterior probability that $h \le h_0$ is $\Pr\{\sum_{i=1}^r \log Z_i \le \log h_0\}$, $Z_1, \ldots, Z_r$ being independent beta variables with parameters $(\alpha_1^*, \beta_1^*), \ldots, (\alpha_r^*, \beta_r^*)$. For small or moderate values of r the numerical integration in (9.14) may be evaluated directly by the methods discussed in Chapter 6, while for larger values it is possible to use a Normal approximation or to improve that by an Edgeworth or saddle-point expansion.

It can be seen that this is conceptually much simpler than the classical contruction of confidence intervals – there is no need to worry about the ordering of the sample space since from the Bayesian viewpoint this is irrelevant. Computationally, this technique will be hard for a large system, especially if the structure is more complicated than a series or parallel system, but in that case approximate methods based on asymptotic normality of the posterior distribution lead to similar results as those obtained for classical confidence limits using similar asymptotics.

The preceding discussion has assumed that it is possible to specify a prior distribution for each component. In some contexts this will not be realistic, but it may still be possible to specify a prior distribution for system reliability. Mastran and Singpurwalla (1978) supposed that a beta or negative log gamma prior density for system reliability is available. In the case of a series system, system reliability is a product of component reliabilities so the task is to find independent prior distributions on the component reliabilities that are consistent with the assumed prior distribution for system reliability. The negative log gamma distribution (i.e. e^{-Y} where Y is gamma) has the property of being closed under product operations, and this is the reason for considering it as an alternative to the conjugate beta prior. Even under this assumption, however, the prior distributions for the individual components are obviously not unique.

Dependent components

This section has been entirely concerned with independent components. When the components are dependent, we may be able to fit one of the models of Chapter 7 to estimate the multivariate reliability function, but the corresponding rise in complexity of the procedure would seem to indicate that there is little hope for an exact method for constructing confidence limits for system reliability. The main alternatives then would be either to use the bootstrap, or else one of the all-purpose approximate methods such as the likelihood ratio technique. Little research has been done on these questions, though it is obviously an area where more is needed.

Example 9.5

The preceding concepts will be illustrated with some numerical examples. Suppose there are three types of component labelled 1, 2, 3, and that $n_i = 10$, 15, 20, $Y_i = 3$, 4, 5 respectively for $i = 1$, 2, 3. Consider two reliability functions: Problem 1 for which $h(p) = p_1 p_2 p_3$, a series system, and Problem 2 for which $h(p) = (3p_1 - 3p_1^2 + p_1^3)p_2(2p_3 - p_3^2)$. The latter is essentially (9.3) with $p_1 = p_2 = p_3$ and $p_5 = p_6$, with p_1, p_4, p_5 respectively relabelled p_1, p_2, p_3.

For Problem 1, the point estimate is $\hat{h} = 0.3850$ with an estimated standard error (by the delta method) of 0.1114. A 95% lower confidence bound by the Normal approximation is 0.2017 and by Easterling's method is 0.2171. A bootstrap based on 1000 replications gave a lower confidence bound of 0.216. The likelihood ratio or profile likelihood function is plotted in Figure 9.2 – a lower confidence limit by this method is 0.218. These estimates are to be compared with the exact lower confidence bound computed by (9.13), which turns out to be 0.15. If we assume vague prior knowledge about the p_i in a Bayesian analysis, and formally take the limiting conjugate prior when $\alpha_i = \beta_i = 0$, then $E\{p_i | Y\} = Y_i / n_i$ and $E\{h(p)|Y\} = 0.385$, which coincides in this case with the maximum likelihood estimator. The exact posterior standard deviation is 0.1406. Finally, a Normal approximation to the posterior distribution of $\log h(p)$ yields a 95% posterior lower bound for $h(p)$ of 0.2.

For Problem 2, the corresponding results are: $\hat{h} = 0.6689$, standard error 0.1130, lower confidence limits 0.483 (Normal approximation), 0.446 (Easterling), 0.462 (bootstrap), 0.471 (likelihood ratio), 0.478 (exact). The profile likelihood is plotted in Figure 9.3. The posterior mean and standard deviation in a Bayesian approach, with the same priors as in Problem 1, are 0.630 and 0.1412. In this case a Normal approximation to the posterior distribution of $h(p)$ yields a 95% posterior lower bound of 0.397, though the accuracy of the Normal approximation is questionable.

In both cases the profile likelihood function is noticeably skewed and this seems to explain the difference between the Normal approximation (which

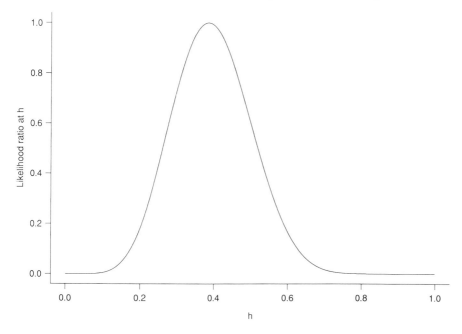

Figure 9.2 Profile likelihood for system reliability h (Problem 1), normalized to have maximum value 1.

assumes symmetry) and the others. In terms of computation, the delta method and Easterling's procedure are very easy to program and virtually instantaneous to run. The bootstrap is also easy to program while both this and the likelihood ratio method take somewhat longer to run than the delta method, but (for this size of problem) are well within the range of a desktop PC. The exact confidence limits are the hardest to compute, but for this example were completed in a few minutes on a SUN 3 workstation.

Our two other examples are of exactly the same structure, but are included to make comparisons with previously published examples. The first is Example 1 of Winterbottom (1984), for which $n_i = 20$, 24, 30, $Y_i =$ 1, 2, 2 respectively for $i = 1$, 2, 3. In this case we compute 90% lower confidence limits for which the exact value (problem 1) turns out to be 0.6767. Winterbottom did not obtain this figure, but gave various bounds which showed that the figure was between 0.6538 and 0.6941. For comparison, we give other values computed by various methods: 0.715, 0.683, 0.709, 0.703 respectively for the Normal, Easterling, bootstrap and likelihood ratio approximations. With Problem 2 (not discussed by Winterbottom) the exact value is 0.7931 and the approximations 0.840, 0.761, 0.830, 0.822.

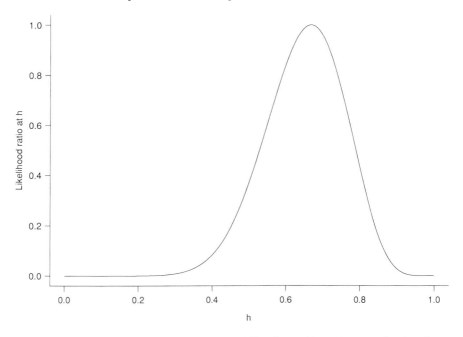

Figure 9.3 Profile likelihood for system reliability h (Problem 2), normalized to have maximum value 1.

For our final example, suppose $n_i = 50$, 50, 50, $Y_i = 1$, 2, 4 respectively for $i = 1, 2, 3$. Here the exact 95% lower bound is 0.7693 by our calculation, which differs slightly from the Lipow–Riley value of 0.767 quoted by Mann and Grubbs (1974) in their Table I. It is not clear why the discrepancy occurs, but a plausible explanation is that our method of ordering states is slightly different from theirs. The normal, Easterling, bootstrap and likelihood ratio approximations are 0.788, 0.766, 0.787 and 0.776. The corresponding values for Problem 2 are 0.8789 (exact) with the approximations 0.9075, 0.867, 0.900 and 0.892.

The foregoing comparisons are not nearly as extensive as the numerical comparisons of Mann and Grubbs (1974), Rice and Moore (1983) and Martz and Duran (1985), but one can make some general observations. The various approximate methods, with the exception of Easterling's, tend to give slightly larger lower confidence limits than the exact limits computed via (9.13). In the past it has been assumed that (9.13) was too complicated to compute exactly, and consequently most of the attention was on approximate methods which were assessed according to how closely they reproduced the result of (9.13) in cases where the latter was computable. Our own studies have suggested that exact computation of (9.13) is not now such a hard problem, at least in relatively small experiments, but have also highlighted some difficulties with the concept. One is the arbitrary ordering of data vectors. A second is that (9.12) is only an inequality, because of the discreteness of the Binomial distribution, and consequently the so-called exact procedure is in fact conservative. A third difficulty is that the minimization over p^*-vectors in (9.13) induces additional conservatism when the true p-vector is different from the one that achieves the minimum. On the other hand, the Bayesian method is based on different principles and avoids these problems, at the cost of requiring a prior distribution. The choice among the various methods depends not only on matters of computational accuracy, but on what the statistician is trying to do.

9.4 MULTI-STATE RELIABILITY THEORY

In multi-state reliability, the state of the system is classified into one of $M+1$ states $S = \{0, 1, 2, \ldots, M\}$. Suppose there are n components labelled $1, \ldots, n$. The state of the ith component is assumed to be in a set S_i, where S_i is a subset of S including at least the states 0 and M. The interpretation is that state 0 represents failure, state M perfect operation, while other states between 0 and M represent various intermediate stages of failure. The subject has been extensively developed in recent years as a logical extension of the theory of coherent systems; see in particular the reviews by Natvig (1985a, 1985b).

For a multi-state system, it is possible to define a structure function in the same way as for a binary system: the structure function is given by

$$\phi = \phi(x), \qquad x_i \in S_i, \; i = 1, \ldots, n$$

where x is a vector of component states $x_1, \ldots, x_n$ and $\phi \in S$ represents the state of the system.

A system is a **multi-state monotone system** (MMS) if its structure function satisfies:

1. $\phi(x)$ is non-decreasing in each argument,
2. $\phi(\mathbf{0}) = 0$ and $\phi(\mathbf{M}) = M$ (here $\mathbf{0}$ and $\mathbf{M}$ are vectors with all components 0 and M, respectively).

A system is a **binary type multi-state monotone system** (BTMMS) if, for each $j \in \{1, \ldots, M\}$, there exists a binary structure function ϕ_j such that

$$\phi(x) \geq j \quad \text{if and only if} \quad \phi_j(x^{(j)}) = 1$$

where $x^{(j)}$ is a vector of $x_i^{(j)}$, where $x_i^{(j)}$ is 1 if $x_i \geq j$, 0 otherwise. In words, a multi-state system is of binary type if, for each j, it can be reduced to a binary system in which the states $0, \ldots, j-1$ are identified with failure and $j, \ldots, M$ with success.

The concepts of minimal path and minimal cut may also be generalized to multi-state systems. In the binary case, a path or a cut is a subset of the components, but we could equally well talk of a path (cut) *vector* x, where x_i is 1 if component i is a member of the path (cut) and 0 otherwise. In a MMS, for $j \in \{1, \dots, M\}$, the vector x is said to be a **minimal path vector to level j** if $\phi(x) \geq j$ and $\phi(y) < j$ whenever $y < x$. Here $y < x$ means $y_i \leq x_i$ for all i with strict inequality for at least one i. Similarly, x is a **minimal cut vector to level j** if $\phi(x) < j$ and $\phi(y) \geq j$ whenever $y > x$.

Example 9.6.
The following example is taken from the very interesting case study of Natvig *et al.* (1986). Two oil rigs R_1 and R_2 are served by generators A_1, A_2 and A_3, each with a capacity of 50 MW. If A_1 fails then A_2 acts as a backup for R_1; however if A_1 is functioning and A_3 fails, the power from A_2 can be diverted to R_2 through an under-sea cable L. The whole system is supervised by a controller U (Figure 9.4). The five components are A_1, A_2, A_3, L and U and the possible states are as follows:

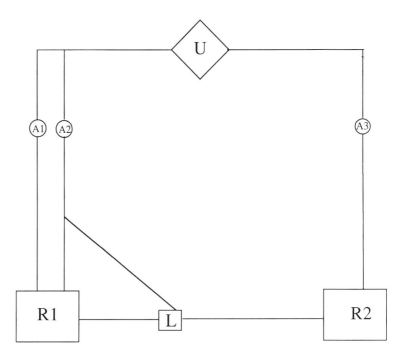

Figure 9.4 Diagram of oil rig system.

For A_1, A_2 and A_3:

      0: the generator cannot supply any power,
      2: the generator can supply 25 MW,
      4: the generator can supply 50 MW.

For L:

      0: the cable cannot transmit any power,
      2: the cable can transmit half the power sent through it.
      4: the cable can transmit all the power sent through it.

For U:

      0: the control unit incorrectly shuts down A_1 and A_3
         without starting up A_2,
      2: the control unit fails to start A_2,
      4: the control unit is working perfectly.

It can be seen that the possible levels of power available to R_2 are 0, 12.5, 25, 37.5 and 50 MW, and these are coded states 0 to 4 respectively. Thus, $M=4$ in this example. R_1 has only states 0, 2, 4. Letting I denote indicator function, the structure functions for R_1 and R_2 are then:

$$\phi_1(U, A_1, A_2) = I(U>0) \min\{A_1 + A_2 I(U=4), 4\},$$
$$\phi_2(U, A_1, L, A_2, A_3) = I(U>0) \min\{A_3 + A_2 I(U=4)I(A_1=4)L/4, 4\}.$$

Note that it is assumed that A_2 only sends power to R_2 when A_1 is working perfectly; it is of course possible to consider a more flexible arrangement whereby A_2's power could be split between the two stations, as well as systems containing a finer classification of states, and both these possibilities are discussed by Natvig et al., but we stick to the simple system here.

Tables 9.1 to 9.4 list the minimal path and minimal cut vectors of ϕ_1 and ϕ_2. As an example of how these tables are calculated, consider the fifth row of Table 9.3. R_2 will receive exactly 37.5 MW if U and A_1 are working perfectly, A_3 is producing half power and both A_2 and L are operating at half capacity. Thus $(4, 4, 2, 2, 2)$ is a path vector $(\phi(4, 4, 2, 2, 2)=3)$, and it is minimal because $\phi(2, 4, 2, 2, 2)$, $\phi(4, 2, 2, 2, 2), \ldots, \phi(4, 4, 2, 2, 0)$ are all less than 3. Similarly, for example, $(4, 4, 2, 2, 0)$ is a minimal cut vector for state 2 of R_2 because, if U and A_1 are both working perfectly, L and A_2 working on half capacity, and A_3 failed, then R_2 is in state 1 (i.e. less than 2) whereas, if any of the component states is improved by even one category, R_2 will be at least in state 2.

Table 9.1: Minimal path vectors of ϕ_1

Levels	U	A_1	A_2
2	2	2	0
2	4	0	2
4	2	4	0
4	4	0	4
4	4	2	2

Analysis of multi-state systems

From the example just discussed, it should be clear that many of the concepts of coherent systems may be extended fairly directly to multi-state systems. Suppose the components are independent and, for component i, the probability that its state is j is known, for each $j \in S_i$. For each path x, one can easily calculate the probability that x is available, i.e. that component i is in a state $\geq x_i$ for each i. Similarly for cuts. The event that the system is in at least state j can be written as a union over minimal paths for state j; this may be used to obtain bounds on the corresponding probability, or exact calculation via the inclusion-exclusion formula. Similarly, the probability that the state of the system is less than j may be bounded or exactly calculated from the minimal cuts for state j. In small systems these calculations may be made tediously but directly by enumerating all possibilities.

Table 9.2 Minimal cut vectors of ϕ_1

Levels	U	A_1	A_2
2	4	0	0
2	2	0	4
2,4	0	4	4
4	4	2	0
4	2	2	4
4	4	0	2

For estimation of system reliability from component data in the multi-state case, all the methods of section 9.3 carry over in principle, the basic distribution for components now being multinomial instead of binomial, though the complications of ordering of the state space, which arise in trying to contruct exact confidence intervals, are even worse in multi-state systems. It would appear that the delta method, which involves easily computed but not necessarily very accurate approximations, and the bootstrap, which is no harder to apply in the multi-state case than in the binary case, are the most

Table 9.3 Minimal path vectors of ϕ_2

Levels	U	A_1	L	A_2	A_3
1	4	4	2	2	0
1,2	2	0	0	0	2
2	4	4	2	4	0
2	4	4	4	2	0
3	4	4	2	2	2
3,4	2	0	0	0	4
4	4	4	2	4	2
4	4	4	4	2	2
3,4	4	4	4	4	0

suitable methods for this purpose; the others are likely to be extremely long-winded computationally.

Bounds and approximations for multi-state systems
Most of the detailed mathematics developed by Natvig and his co-workers has been concerned with bounds for the availability of a multi-state system, under stochastic process assumptions about individual components. The performance process for the ith component is defined to be a stochastic process $\{X_i(t), t \geq 0\}$, taking values in S_i, where the interpretation is that $X_i(t)$ is the state of component i at time t. The vector $\{X(t) = (X_1(t), \ldots, X_n(t)), t \geq 0\}$ is the **joint performance process of the components**, and $\{\phi(X(t)), t \geq 0\}$ is the **performance process of the system**. For a time interval I, the availability $h_\phi^{j(I)}$ and unavailability $g_\phi^{j(I)}$ of the system to level j in the interval I denote the probability that the

Table 9.4 Minimal cut vectors of ϕ_2

Levels	U	A_1	L	A_2	A_3
1,2	4	4	4	0	0
1,2,3,4	0	4	4	4	4
1,2	4	4	0	4	0
1,2	4	2	4	4	0
1,2	2	4	4	4	0
2	4	4	2	2	0
3,4	4	4	4	0	2
3,4	4	4	0	4	2
3,4	4	2	4	4	2
3,4	2	4	4	4	2
3,4	4	4	2	4	0
3,4	4	4	4	2	0
4	4	4	2	2	2

system is in state $\geq j$ (for availability) or $<j$ (for unavailability) throughout the interval I. Thus

$$h_\phi^{j(I)} = \Pr\{\phi(X(s)) \geq j, s \in I\}, \qquad g_\phi^{j(I)} = \Pr\{\phi(X(s)) < j, s \in I\}.$$

Note that $h_\phi^{j(I)} + g_\phi^{j(I)} = 1$ when I consists of one point, but in general $h_\phi^{j(I)} + g_\phi^{j(I)} \leq 1$.

It is assumed that the performance processes for components are independent, i.e. that the vectors $(X_i(t_1), \ldots, X_i(t_m))$ are independent for $i = 1, \ldots, n$, for any fixed set of times $t_1, \ldots, t_m$, and also that the individual performance processes are **associated**, defined to mean that, for each i and $t_1, \ldots, t_m$, the vector $(X_i(t_1), \ldots, X_i(t_m))$ is associated in the sense described in section 9.2. Under these assumptions Funnemark and Natvig (1985) and Natvig (1986) were able to derive bounds for the functions $h_\phi^{j(I)}$ and $g_\phi^{j(I)}$ in terms of the individual component reliabilities. These results may be considered to be the generalization to multi-state systems and stochastic processes of the inequalities (9.9).

Continuation of example

We return to the oil rig example considered earlier. Natvig *et al.* (1986) modelled each component performance process as a three-state continuous-time homogeneous Markov chain. Since the number of states is small, it is relatively straightforward to write down the $p_i^{j(I)}, q_i^{j(I)}$ expressions exactly (it requires diagonalising a 3×3 matrix of instantaneous transition rates and using the exponential matrix form of the transition matrix). Making educated guesses about the transition rates, they were able to produce a table of these values. They also stated a condition for a continuous-time Markov process to be associated, and checked that their assumed parameters satisfied this. Hence, using the minimal path and cut representations of Tables 9.1–4, they obtained bounds on system availability and unavailability for several different intervals I. The bounds were in some cases very wide (one of their quoted calculations for h had a lower bound of 0.0002 and an upper bound of 0.9995), but in cases where I was comparatively short they were much tighter than this. For further details we refer to their paper.

Conclusions

Multi-state systems are a logical extension of binary coherent systems, and it can be seen that many of the important concepts carry over. However, the increase in computational complexity is quite dramatic — in the example discussed here, there are only 3 states for each component and 5 for the system, yet the calculations are much more complicated than for a binary system. The development of sharper inequalities and more efficient computational algorithms is clearly an important line of research, but for practical implementation a more pragmatic course of action may simply be to use simulation for the system under study.

9.5 LOAD-SHARING SYSTEMS: THE DANIELS MODEL

We now turn to systems characterized by load-sharing over the components. There is an overlap here with multi-state systems: each component may be classified into one of a number of states according to its strength or its failure time under some particular load, and the probabilities of different states of the system (strengths or failure times) computed by the methods of multi-state systems. However, the concepts described in section 9.4 are geared more to exact calculations or bounds in small systems, whereas our interest in this and the following sections is more concerned with large systems in which there are possibly thousands of components. In such situations exact calculation becomes impossible, but there are numerous specific models for which approximate or asymptotic results are available. Our purpose, then, is to survey some of the more important of these.

The simplest model is one in which there are n components, each having a strength (a continuous and positive random variable) and sharing the applied load equally. The strength of the system is then the largest load that can be applied to the system without every component failing. This model was originally studied by Peirce (1926), but most of the fundamental mathematical work on it originates with Daniels (1945), and it is for that reason that Daniels' is the name most closely associated with it. This and related models were originally developed in the context of textiles and are today often applied to composite materials so we shall generally refer to the components as fibres and the system as a bundle of fibres, though it should be emphasized that the applications are not restricted to fibrous systems.

First we consider exact formulae for the distribution of bundle strength. A small convention which should be noted is that we always measure strength or load on the bundle in terms of load per fibre: apart from the fact that this is the logical definition from the engineering point of view (engineers usually quote material strength in terms of failure stress rather than failure load) it also simplifies notation in the mathematical formulae.

Let $F(x)$, $x \geq 0$ denote the distribution function for individual fibres, i.e. $F(x)$ is the probability that a fibre fails when subjected to load x. We shall also denote by $G_n(x)$ the probability of failure for a bundle of n fibres subjected to load (per fibre) x, and we assume the individual fibre strengths are independent. Consider the case $n=2$. The bundle fails under load x if one of the fibres has strength less than x and the other less than $2x$. Hence

$$G_2(x) = 2F(x)F(2x) - F(x)^2. \tag{9.15}$$

This comes from $\Pr\{A \cup B\} = \Pr\{A\} + \Pr\{B\} - \Pr\{A \cap B\}$ where A is the event that the first fibre fails under load x and the second under load $2x$, and B is the event that the second fibre fails under load x and the first under $2x$. Both $\Pr\{A\}$ and $\Pr\{B\}$ are equal to $F(x)F(2x)$, and $\Pr\{A \cap B\}$ is the probability that both fibres fail under load x, i.e. $F(x)^2$.

For $n = 3$, the bundle will fail under any of the following scenarios: (a) one fibre has strength less than x, a second between x and $3x/2$, the third between $3x/2$ and $3x$; (b) one fibre has strength less than x, both the others between x and $3x/2$; (c) two fibres have strength less than x, the third between x and $3x$; (d) all three fibres have strength less than x. Putting these probabilities together yields

$$G_3(x) = 6F(x)(F(3x/2) - F(x))(F(3x) - F(3x/2)) + 3F(x)(F(3x/2) - F(x))^2$$
$$+ 3F(x)^2(F(3x) - F(x)) + F(x)^3. \tag{9.16}$$

It is obvious that this sort of calculation will quickly become prohibitive as n increases. Fortunately, there are a number of recursive formulae which greatly simplify the calculation. We shall outline one of these here, which is particularly appealing because it is a direct application of the inclusion-exclusion formula (9.8).

If the bundle is to fail, then there must be some group of fibres which fail under the original load x, and the rest of the bundle must fail under the load redistributed from that group. For $i = 1, \dots, n$, let A_i denote the event that the bundle fails and fibre i has strength less than x. Since there must be at least one such initial failure, the event of bundle failure may be written $\bigcup_{i=1}^n A_i$. By the inclusion-exclusion formula (9.8), this probability may be written in terms of events of the form $\bigcap_{i \in I} A_i$ for non-empty subsets I of $\{1, \dots, n\}$. However, if I has r members,

$$\Pr\left\{ \bigcap_{i \in I} A_i \right\} = F^r(x)G_{n-r}(nx/(n-r)).$$

This formula reflects the fact that, if all the fibres in I fail under their initial load, that leaves $n - r$ fibres holding a load per fibre of $nx/(n-r)$. Grouping all subsets I of the same size in (9.8), we then deduce

$$G_n(x) = \sum_{r=1}^n \binom{n}{r}(-1)^{r+1}F^r(x)G_{n-r}(nx/(n-r)). \tag{9.17}$$

The reader is left to verify that (9.17) indeed reduces to (9.15) or (9.16) in the cases $n = 2, 3$.

The idea of using the inclusion-exclusion formula in this way is apparently due to Suh, Bhattacharyya and Grandage (1970), though recursive formulae of this type go back to Daniels' original 1945 paper. Formula (9.17) is fast and easy to program, but becomes numerically unstable for large n (greater than about 40) – this essentially arises because it is a long series of terms of alternating sign. McCartney and Smith (1983) derived an alternative recursive formula which gives much better numerical results for larger n. We have chosen to concentrate on (9.17) here because the basic idea is easily generalized to other models (section 9.10).

Approximations for large bundles
For large values of n it is not necessary to calculate the distribution function of bundle strength exactly, because there are some very good approximations. To introduce these, it is convenient to adopt a slight change of perspective, to a controlled strain approach. Imagine the fibres stretched in tension between two rigid clamps, which are then gradually moved apart. Thus it is the extension in the fibres, or strain, which is under experimental control. For elastic fibres, Hooke's law specifies a linear relationship between stress and strain, so the load on each fibre increases linearly with strain. Thus, controlling strain is equivalent to controlling stress per fibre, as opposed to an experiment in which it is the total load on the system that is under control. Now, suppose the n fibres fail randomly and let F_n denote the empirical distribution function of the fibre strengths. That is, $nF_n(x)$ is the number of failed fibres when the load per fibre is x. Hence the actual load supported by the system at this point is $x(1-F_n(x))$, still measured in terms of load per fibre of the original bundle. Figure 9.5 (based on a simulation with 40 fibres) shows a typical behaviour of this function as x increases. Also shown on the figure is $x(1-F(x))$, the limiting form of this function as $n\to\infty$. The quantity of interest to us is the maximum load the system can bear, i.e. $\max_{x>0}\{x(1-F_n(x))\}$.

Now consider what happens for large n. For fixed x, $nF_n(x)$ has a Binomial distribution with mean $nF(x)$ and variance $nF(x)(1-F(x))$. Thus $Q_n(x)=n^{1/2}\{x(1-F_n(x))-x(1-F(x))\}$ has mean 0 and variance $x^2F(x)(1-F(x))$, and by the Central Limit Theorem has an asymptotically Normal distribution as $n\to\infty$. In fact, though this is only really needed much later, the stochastic process $Q_n(x)$ converges to a limiting Gaussian stochastic process $Q(x)$ with mean 0 and covariance function

$$\mathrm{cov}\{Q(x), Q(y)\}=xyF(x)(1-F(y)), \; x\le y.$$

We shall assume that the function $x(1-F(x))$ has a unique maximum $M=x^*(1-F(x^*))$ achieved at $x=x^*$, and also that the second derivative of

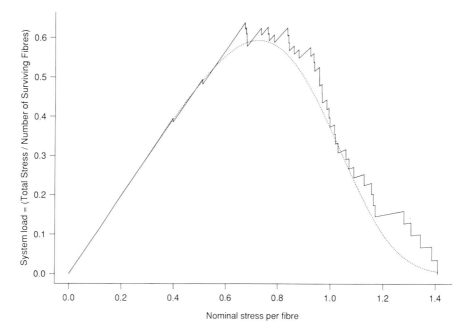

Figure 9.5 A simulation of Daniels' failure model. The solid line is $x(1\text{-}F_n(x))$ where F_n denotes the empirical distribution function from a sample of n fibre strengths; here $n=40$. The limiting curve $x(1\text{-}F(x))$ is shown by the dotted line, where $F(x)$ is the distribution function of a Weibull random variable with shape parameter 5.

$x(1-F(x))$ is <0 at this point. To fix some useful notation, let $S^2 = (x^*)^2 F(x^*)(1-F(x^*))$ and $B^3 = (x^*)^4 \{F'(x^*)\}^2/\{2F'(x^*)+x^*F''(x^*)\}$. A first approximation to the distribution of bundle strength is to assume that the maximum of $x(1-F_n(x))$ is achieved at $x=x^*$. Under this approximation, it has an approximate Normal distribution with mean M and variance $n^{-1}S^2$. Thus we have

First approximation

$$F_n(x) \approx \Phi\{n^{1/2}(x-M)/S\}$$

where Φ is the standard Normal distribution function. The rigorous derivation of this was essentially the main conclusion of Daniels (1945), though Phoenix and Taylor (1973) gave a shorter argument based on weak convergence theory to tighten up the reasoning just given, whereas Daniels' original paper used quite different methods.

This is not a very good approximation. In classical central limit theory, the error in the normal approximation is of $O(n^{-1/2})$, whereas for this result the corresponding error is of $O(n^{-1/6})$. Daniels (1974a,b), Barbour (1975) and Smith (1982) all investigated an improved approximation which allows for the random perturbations of $Q_n(x)$ in the neighbourhood of $x = x^*$. This leads to the following result.

Second approximation

$$F_n(x) \approx \Phi\{n^{1/2}(x - M - \lambda n^{-2/3}B)/S\}.$$

Here λ is an absolute constant whose numerical value is about 0.996. In words, the mean is 'corrected' by adding on a term $\lambda n^{-2/3}B$. This reduces the error in the approximation to $O(n^{-1/3})$.

A further refinement by Barbour (1981) led to the following.

Third approximation

$$F_n(x) \approx \Phi[n^{1/2}(x - M - \lambda n^{-2/3}B)/\{S^2 + \phi n^{-1/3}B^2\}^{1/2}]$$

Here ϕ is another absolute constant whose numerical value is about -0.317. Barbour's approximation corrects for the variance as well as the mean and finally obtains an approximation with error $O(n^{-1/2})$.

Example 9.7

Suppose $F(x) = 1 - \exp(-x^\alpha)$, a two-parameter Weibull distribution with scale parameter 1 and shape parameter α. Of course, by rescaling the following reasoning applies to any scale parameter. In this case it is readily checked that $x^* = \alpha^{-1/\alpha}$, $M = \alpha^{-1/\alpha}e^{-1/\alpha}$, $S/M = (e^{1/\alpha} - 1)^{1/2}$, $B/M = \alpha^{-1/3}e^{2/(3\alpha)}$. Thus with $\alpha = 5$, say, we have $M = 0.5934$, $S = 0.2792$, $B = 0.3965$. Evaluating the approximation for $n = 30$, we have

First approximation: Mean bundle strength   standard deviation
                         $= 0.5934$, $= 0.05097$.
Second approximation: Mean bundle strength   standard deviation
                         $= 0.6343$, $= 0.05097$.
Third approximation: Mean bundle strength   standard deviation
                         $= 0.6343$, $= 0.04543$.

Similarly, for $n = 500$ we have

First approximation: Mean bundle strength   standard deviation
                         $= 0.5934$, $= 0.01251$.
Second approximation: Mean bundle strength   standard deviation
                         $= 0.5997$, $= 0.01251$.

Third approximation: Mean bundle strength   standard deviation
$$= 0.5997, \qquad\qquad\qquad = 0.01197.$$

These figures suggest (especially for $n = 30$) that the improvement from the first to the second approximation is important, since the change in estimated mean is comparable with the standard deviation. However, the improvement from the second to the third approximation is less important. This rough conclusion was confirmed in a detailed numerical study carried out by McCartney and Smith (1983), though they concluded that the third approximation is nevertheless a valuable refinement in the lower tail of the distribution, where sensitive approximations are particularly important.

In conclusion to this section, it may be noted that approximations of the form developed by Daniels and others have proved useful in a variety of probabilistic approximation problems. Indeed the paper Daniels (1974b) was primarily about an epidemic model for which the same approximation problem arises, and the more recent work of Groeneboom (1989) was motivated by the problem of estimating the mode of a distribution. Other developments are Daniels and Skyrme (1985), who developed asymptotic results for the joint distribution of $\max_x\{x(1 - F_n(x))\}$ and the value of x at which the maximum is attained, and Daniels (1989) who extended the result to a wider class of processes. We omit details of these results, contenting ourselves with the remark that they go a long way beyond the original rather simple model which motivated the theory.

9.6* EXTENSIONS OF THE DANIELS MODEL

Phoenix and Taylor (1973) proposed an extension of the Daniels model which allows for extra sources of variability, in particular the possibility of random slack in the fibres or effects such as random change from elastic to plastic behaviour. The basic approach is again the controlled strain approach described in the previous section, where this time we do not immediately transform the independent variable back from strain to stress. The mathematics in this section makes much more explicit use of the stochastic process features of the problem, and consequently makes the section somewhat harder than those which precede it.

Suppose the load borne by the ith fibre under strain t is a random function of the form

$$y_i(t) = \begin{cases} q(t, X_i), & t \le T_i, \\ 0, & t > T_i, \end{cases}$$

where q is a function depending on a variable X_i which differs from fibre to fibre, and T_i is the failure strain of the fibre. The total load on the bundle at strain t (measured in load per fibre) is

$$Y_n(t) = n^{-1} \sum_{i=1}^{n} y_i(t).$$

It is assumed that the fibres are independent and a priori identical, in other words that the pairs (X_i, T_i) are independent and identically distributed for $i = 1, \ldots, n$.

Suppose $\mu(t) = E\{y_i(t)\}$ and $\Gamma(s, t) = \mathrm{Cov}\{y_i(s), y_i(t)\}$. Then, for fixed t, $Y_n(t)$ has mean $\mu(t)$ and variance $n^{-1}\Gamma(t, t)$. Following the same reasoning as in the previous section, suppose $\mu(t)$ achieves a unique maximum M at $t = t^*$, and let $S^2 = \Gamma(t^*, t^*)$. We are led to the following.

First appoximation:
Bundle strength has approximately a Normal distribution with mean M and variance S^2/N.

Phoenix and Taylor (1973) gave rigorous conditions on the distribution of (X_i, T_i) and the function q for the distribution of $n^{1/2}[\{\sup_{t>0} Y_n(t)\} - M]/S$ to converge to standard Normal as $n \to \infty$. They also gave a number of examples of problems fitting within this framework. We describe two of them here.

Example 9.8: Random slack
In many cases, the fibres are not all of exactly the same length, so that when they are initially stretched some or all of them are slightly slack. The amount of slack in each fibre may itself be considered a random variable.

Suppose fibre i has Young's modulus C and random slack X_i. Thus

$$q(t, X_i) = C(t - X_i)_+$$

where y_+ is y if $y > 0$, 0 otherwise. The failure strain may be written in the form $T_i = X_i + Z_i$ where X_i and Z_i are independent, Z_i being the actual strain in the fibre at failure when the slack is taken into account. Then

$$y_i(t) = \begin{cases} C(t - X_i)_+, & t \le X_i + Y_i, \\ 0, & t > X_i + Y_i. \end{cases}$$

The functions μ and Γ may be calculated as

$$\mu(t) = \int_0^t C(t - u)\{1 - H(t - u)\} \, dG(u),$$

$$\Gamma(s, t) = \int_0^s C(s - u)C(t - u)\{1 - H(t - u)\} \, dG(u) - \mu(s)\mu(t)(s \le t)$$

where G and H are the distribution functions for X_i and Z_i respectively. It can be shown that $\sup_{t>0}\mu(t)\le\sup_{t>0}\{q(t)(1-H(t))\}$, which shows that imperfect loading has the effect of decreasing mean bundle strength.

Example 9.9: Elastic-plastic behaviour
Another kind of model to which this scenario applies is one in which a fibre behaves elastically up to a certain (random) point, and then plastically up to its random failure strain. This can be modelled by

$$q(t, X_i)=\begin{cases}Ct, & t<X_i,\\CX_i, & t\ge X_i,\end{cases}$$

where X_i is the transition point from elastic to plastic behaviour. Again it is reasonable to assume $T_i=X_i+Z_i$ for independent X_i and Z_i with distribution functions G and H, where Z_i is the increment in strain during the plastic phase up to failure. Then

$$\mu(t)=C\left[t\{1-G(t)\}+\int_0^t u\{1-H(t-u)\}\,dG(u)\right],$$

$$\Gamma(s, t)=C^2\left[st\{1-G(t)\}+s\int_s^t u\{1-H(t-u)\}\,dG(u)\right.$$

$$\left.+\int_0^s u^2\{1-H(t-u)\}\,dG(u)\right]-\mu(s)\mu(t), s<t.$$

In this case it can be shown that $\sup_{t>0}\mu(t)\ge\sup_{t<0}\{Ct(1-G(t))\}$, so a bundle with elastic-plastic behaviour is stronger than one with the same behaviour up to the point of elastic failure but no plastic phase.

Improved approximations
Daniels (1989) extended the work of Daniels (1974a,b) and Daniels and Skyrme (1985) to develop an improved approximation for a general class of random processes which can be approximated by a Gaussian process. Recall that the first approximation to the strength of a bundle of n fibres is Normal with mean $\mu(t^*)$ and variance $n^{-1}\Gamma(t^*,t^*)$ where t^* maximizes $\mu(t)$. The improved approximation is Normal with the same variance but mean now replaced by $\mu(t^*)+\lambda n^{-2/3}A^{2/3}\{-\mu''(t^*)\}^{-1/3}$ where $\lambda\approx0.996$ is as before and A is a constant defined as follows. Let $\Gamma'=\partial\Gamma(s, t)/\partial s$, $\dot{\Gamma}=\partial\Gamma(s, t)/\partial t$ each evaluated as $s=t=t^*$. Then $A=\Gamma'-\dot{\Gamma}>0$. Although this result is superficially simple its derivation involves different Gaussian process approximations for the cases when Γ' and $\dot{\Gamma}$ have the same sign or opposite

signs. Daniels also derived limiting joint and conditional distributions for the maximum of the process and the value of t at which it is achieved.

9.7* TIME TO FAILURE

We now consider time-dependent models in which the total load on the bundle is either held constant or subject to some well-defined (e.g. linearly increasing or cyclic) variation, and the variable of interest is failure time. Models for this were considered in a series of papers by Coleman (1956), and given a rigorous mathematical development by Phoenix (1978, 1979). We follow Phoenix here, and refer the reader to his papers for more detailed description of the background. The assumption throughout is equal load-sharing, except for the random slack problem which is briefly discussed at the end of this section.

Since the load on individual fibres varies with the number of failed fibres, even when the total load on the system remains constant, we must consider situations of varying load on a single fibre. Phoenix (1978), following Coleman, defined the model

$$\Pr\{T \le t | L(s), s \le t\} = G\left\{\int_0^t \kappa(L(s))\,ds\right\} \qquad (9.18)$$

for the time to failure T of a single fibre under load history $\{L(s), s \ge 0\}$ (i.e. $L(s)$ is the load on the fibre at time s), G is a distribution function possessing (at least) two finite moments and κ is a functional of L called the **breakdown rule**. Intuitively, $\kappa(L(s))\,ds$ represents the instantaneous damage or 'breakdown' in the fibre in the interval $(s, s+ds)$. The breakdown rule is generally taken to be one of two forms.

1. Power-law breakdown

$$\kappa(x) = (x/L_0)^\beta \quad \text{for} \quad L_0 > 0, \beta \ge 1. \qquad (9.19)$$

2. Exponential breakdown

$$\kappa(x) = \alpha e^{\zeta x} \quad \text{for} \quad \alpha > 0, \zeta > 0.$$

It has generally been assumed that the exponential breakdown rule is closer to the predictions of statistical mechanical theory but an appendix to Phoenix and Tierney (1983) challenged this, arguing that both rules are equally valid from this point of view and preferring the power-law breakdown rule for ease of mathematical modelling. Here we shall assume power-law breakdown, though the case of constant total load is also handled by the theory of Phoenix (1978) for a general breakdown rule subject to natural monotonicity and integrability conditions. Note the restriction $\beta \geq 1$; in applications this is usually satisfied with something to spare. The distribution function G is less critical to the analysis, but it is often taken to be of Weibull form $G(x) = 1 - \exp(-x^{\eta})$ where η is typically small, often less than 1. For an example of the statistical problems of determining such parameters the reader is reminded of section 4.10, in which just this model was fitted to Kevlar data with estimates $\hat{\beta} = 23$, $\hat{\eta} = 0.79$. The reader is also referred back to section 4.6, where some similar classes of models were derived from fracture mechanics considerations.

Now we describe the mathematical analysis. Suppose we have a bundle of n fibres and the load on the bundle at time t is $L(t)$. As in previous sections, this is measured in load per fibre, so the total load on the bundle is $nL(t)$. Also let $L_i(t) = nL(t)/(n-i)$, the actual load on each surviving fibre at time t if at that time i have failed and the rest survive.

Let $T_0 = 0$, T_1 denote the time of the first fibre failure, T_2 the time of the second, and so on up to T_n which is the time to failure of the whole bundle. A key result is the following:

Lemma:
Define $W_0 \leq W_1 \leq \cdots \leq W_n$ by $W_0 = 0$,

$$W_{i+1} - W_i = \int_{T_i}^{T_{i+1}} \kappa(L_i(s))\, ds, \, i = 0 \ldots n-1.$$

Then $W_1 \leq \cdots \leq W_n$ are distributed the same as the order statistics of a sample of size n from the distribution function G.

Using this result one can express the failure times $T_1, \ldots, T_n$ in terms of the order statistics $W_1, \ldots, W_n$. The advantage of using power-law breakdown rule for this is that it allows us to deal with a general load function L. This is because under (9.19)

$$\int_{T_i}^{T_{i+1}} \kappa(L_i(s))\, ds = (1 - i/n)^{-\beta} \int_{T_i}^{T_{i+1}} (L(s)/L_0)^{\beta}\, ds$$

and hence

$$S_n = \int_0^{T_n} (L(s)/L_0)^{\beta}\, ds = \sum_{i=1}^{n} \{\phi(i/n) - \phi((i-1)/n)\} W_i \tag{9.20}$$

where $\phi(x) = (1-x)^{\beta}$.

The right hand side of (9.20) is a linear combination of order statistics, a subject for which a great deal of asymptotic theory was developed in the 1960s and 1970s. Phoenix (1978) quoted results of Stigler (1973) to deduce that, for large n, S_n has an approximately Normal distribution with mean μ and variance σ^2/n, where

$$\mu = \int_0^\infty \phi(G(y))\,dy,$$

$$\sigma^2 = \int_0^\infty \int_0^\infty \phi'(G(u))\phi'(G(v))\Gamma(u,v)\,du\,dv,$$

where $\Gamma(u,v) = \Gamma(v,u) = G(u)\{1 - G(v)\}$ for $u \le v$. This assumes finiteness of μ and σ^2 as just defined, and other more technical assumptions that we shall not go into.

In cases where $L(s)^\beta$ is analytically integrable, the asymptotic distribution of S_n from (9.20) may be used directly to compute that of T_n. In particular, if $L(s)$ is a constant L, we have

$$T_n = S_n(L/L_0)^{-\beta}$$

which we should note, recalling the survey of models in Chapter 4, is another accelerated life model in the dependence of T_n on L.

Phoenix also developed the cases of a linearly increasing load $L(t) = L_1 t$ and cyclic loading $L(t) = L_2(t - mb)$ where L_2 is a periodic function of period b and m is the number of complete cycles up to time t, and worked out a number of specific examples of these formulae. One interesting conclusion is that, on a load-per-fibre basis, the mean failure time of a bundle is often an order of magnitude smaller than that of a single fibre under the same load, but this can be restored by a modest decrease in load, and the benefit is a great reduction in the variance of failure time.

Tierney (1981) developed an elaborate theory to obtain an extension of these results for random slack. For mathematical reasons he found it necessary to assume the exponential breakdown rule – this may act as a deterrent to using his results, though over narrow ranges of load it is hard to distinguish between a power-law breakdown with high β and an exponential rule.

9.8* A MORE GENERAL MODEL

Recalling section 4.6, a natural parametric model for failure time under variable load, that takes into account both the initial variability of crack size and long-term crack growth, is

$$\Pr\{T \le t \mid L(s), s \le t\} = \sup_{\tau \le t} G\left[\left\{K_1 L^{\beta - 2}(\tau) + K_2 \int_0^\tau L^\beta(s)\,ds\right\}\right] \quad (9.21)$$

where G is a (usually Weibull) distribution function. Equation (9.21) encompasses both the Weibull distribution for static strength when $t=0$ and the model (9.18) (with power-law breakdown rule) when t is large and the load relatively small.

Phoenix (1979) proposed a general model encompassing this:

$$\Pr\{T \leq t | L(s), s \leq t\} = \sup_{\tau \leq t} \Psi\left\{L(t), \int_0^\tau \kappa(L(s))\,ds\right\}$$

where κ is a breakdown function as before and $\Psi(x, z)$ is increasing and jointly continuous in $x \geq 0, z \geq 0$ with $\Psi(0, 0) = 0$, $\Psi(x, z) < 1$ for all finite (x, z) and $\lim_{x \to \infty} \Psi(x, z) = 1$ for all z, $\lim_{z \to \infty} \Psi(x, z) = 1$ for all x. This is a more general model encompassing both the Daniels and Coleman models as special cases.

Consider a hypothetically infinite bundle with load per fibre L. Let $F(x) = \Psi(x, 0)$ denote the distribution function of initial failures. At time 0, a fraction $y_1 = F(L)$ of the fibres will fail under their original load. The load per fibre immediately jumps to $L/(1 - y_1)$ and the fraction of failed fibres increases instantaneously to $y_2 = F(L/(1 - y_1))$. Continuing the same reasoning, the fraction of failed fibres at the mth load increment is $y_m = F(L/(1 - y_{m-1}))$. Then either $y_m \to 1$, meaning the whole bundle collapses at time 0, or $y_m \to y(0)$ where $y(0)$ is a stable solution of the equation $y = F(L/(1 - y))$. Assume we are in the latter case, i.e. $y(0) < 1$. It can be shown that this case holds whenever $L < L^*$ where L^* is the critical failure load for the Daniels bundle, i.e. maximum of $x(1 - F(x)), x > 0$. By continuing this reasoning for $t > 0$, a stable solution of the system at time t will satisfy

$$y(t) = \Psi\left\{L/(1 - y(t)), \int_0^t \kappa(L/(1 - y(\tau)))\,d\tau\right\} \tag{9.22}$$

so long as $y(t) < 1$. Let $\Psi^{-1}(x, y)$ denote the inverse function of $\Psi(x, z)$ in z for fixed x. That is (under continuity assumptions) $\Psi^{-1}(x, y) = z$ if and only if $\Psi(x, z) = y$. Then we can solve (9.22) to give, over a range of t,

$$\int_0^t \kappa(L/(1 - y(\tau)))\,d\tau = g(y(t)) \tag{9.23}$$

where

$$g(y) = \Psi^{-1}(L/(1 - y), y), \qquad 0 \leq y < 1.$$

In general it can be shown that there exist $0 \leq y_1^\# \leq y_2^\# \leq y_3^\#$ such that $g(y)$ is zero on $(0, y_1^\#)$ (this region will be non-empty whenever $F(L/(1 - y)) < y$), g is increasing on $(y_1^\#, y_2^\#)$, decreasing on $(y_2^\#, y_3^\#)$ and zero for $y > y_3^\#$. Differentiating (9.23) we have

$$dt = \phi(y(t)) \, dg(y(t)), \qquad (y_1^\# \le y < y_2^\#),$$

where

$$\phi(y) = 1/\kappa(L/(1-y)) \quad \text{for} \quad 0 \le y < 1.$$

The solution for the time $t(y)$ needed to achieve a fraction of failed fibres y, assuming it does not happen immediately, is given by

$$t(y) = \int_{y_1^\#}^{y} \phi(z) \, dg(z) + g(y_1^\#) \phi(y_1^\#), \qquad y_1^\# \le y < y_2^\#. \tag{9.24}$$

This reasoning, however, is only valid until y reaches $y_2^\#$. After that the left hand side of (9.23) continues to increase and the right hand side does not. In fact this is the point at which the bundle fails, i.e. y jumps instantaneously to 1. Hence the time to failure $t(1)$ is given by substituting $y = y_2^\#$ in (9.24). Taking this and integrating by parts, we then have

$$t(y) = -\int_0^1 g^\#(\min(z, y)) \, d\phi(z), \qquad L > 0, 0 < y \le 1, \tag{9.25}$$

where $g^\#(y) = \sup\{g(\tau) : 0 \le \tau \le y\}$ and we use the fact that $\phi(y) \to 0$ as $y \to 1$.
   As an example, consider (9.21) with

$$G(y) = 1 - \exp\{(y/y^*)^\eta\}, \qquad y^* > 0, \eta > 0. \tag{9.26}$$

Then it can be seen that

$$g(y) = y^*(-\log y)^{1/\eta} - y^* K_1 l^{\beta-2}(1-y)^{2-\beta}, \qquad y_1^\# < y < y_3^\#, \tag{9.27}$$

with $g = 0$ outside $(y_1^\#, y_3^\#)$. Furthermore

$$\phi(y) = L^{-\beta}(1-y)^\beta \tag{9.28}$$

Moreover equation (9.25) with $y = 1$ reduces to

$$t(1) = -\int_{y_1^\#}^{y_2^\#} g(z) \, d\phi(z) - g(y_2^\#) \int_{y_2^\#}^1 d\phi(z). \tag{9.29}$$

Combining (9.27)–(9.29), with numerical solution for $y_1^\#$ and $y_2^\#$, and numerical integration in (9.29), leads to an explicit solution for the limiting failure time $t(1)$ in an ideal bundle.

An independent investigation of the same problem was by Kelly and McCartney (1981). Their starting point was the need to obtain a model for failure by stress corrosion in bundles of glass fibres. For single fibres they derived a model equivalent to the combination of (9.21) and (9.26). For bundles, although their line of reasoning was quite different from Phoenix's their conclusion was essentially the same series of equations for $t(1)$. They compared the results with experimental data on E-glass fibres immersed in water.

Phoenix (1979) went on to consider the probability distribution for the time to failure in a finite bundle, seeking similar asymptotic results to those quoted earlier for the Daniels and Coleman models. The results are too complicated to quote in full, and are different for $L<L^*$, $L=L^*$ and $L>L^*$ where L^* is the critical load described earlier. We quote only the result for $L<L^*$. In this case the asymptotic distribution of failure time T_n, for a bundle of n fibres, is normal with mean $t(1)$ and variance $\sigma^2(1)/n$, where $\sigma^2(1)$ is defined by

$$\sigma^2(1) = \int_0^1 \int_0^1 \phi'(s)\phi'(t)\Gamma(\min(s, y_2^\#), \min(t, y_2^\#)) \, ds \, dt,$$

the covariance function Γ being defined by

$$\Gamma(s, t) = (\min(s, t) - st)g^{\hat{}}(s)g^{\hat{}}(t) \, ds \, dt,$$
$$g^{\hat{}}(t) = \{\partial \Psi^{-1}(x, t)/\partial t\}|_{x = L/(1-t)}.$$

9.9* LOCAL LOAD-SHARING

The models in Sections 9.5–8 do not allow for any form of physical interaction among the fibres. For example, even in a loose system of fibres one might expect some twisting of the fibres, or frictional force between neighbouring fibres, which will create local stress concentrations and so invalidate the literal equal load-sharing concept. The situation is far worse in a composite, in which the fibres consist of some strong brittle material (typically carbon, glass or a polymer) embedded in a ductile matrix such as epoxy resin. In such materials the fibres continue to bear nearly all the load, but the matrix has an important influence in localizing the stress concentrations. This leads to a quite different class of load-sharing models.

The classical model for this kind of system, whose origins are in some of the early papers on the subject such as Rosen (1964), Zweben and Rosen (1970)

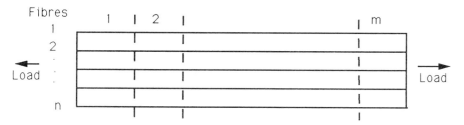

Figure 9.6 The chain of bundles model.

and Scop and Argon (1969), is the **chain of bundles models** (Figure 9.6). Under this model, the material, which is assumed to contain n fibres of length L, is broken up into short bundles of length δ, where δ is the ineffective length. The interpretation is that stress concentrations arising as a result, say, of a failure of one fibre at a particular point, are confined to a small neighbourhood of that point, and in particular are confined within a distance δ along the length of the fibres. The value of δ depends on a number of physical properties of the materials, including the elastic modulus of the fibres, the shear modulus of the fibre-matrix interface and the fibre volume fraction, but in modern composite materials δ is typically very small, or the order of 5–10 fibre diameters. Hence the first simplification is to treat the system as, in effect, a series system of short bundles which are (another obvious simplification) assumed statistically independent. If $F_{n,L}$ denotes the distribution function of either static strength or failure time for a system of n fibres of length L, then this leads to the approximation

$$F_{n,L}(x) = \{F_{n,\delta}(x)\}^{L/\delta} \tag{9.30}$$

It is possible to proceed from (9.30) by treating the individual bundles of length δ as one of the equal load-sharing models discussed in the preceding sections. In particular, this model was studied by Smith (1982) for the classical Daniels model, and by Borges (1983) for the special case of the Coleman model (section 9.7) in which the distribution function G, for the failure time of an individual fibre under constant load, is assumed to be of exponential form. However, this is something of a sideline to the main thrust of the theory, which has been to concentrate on models assuming local stress concentrations within the short bundles of length δ.

Following a series of papers by Harlow and Phoenix (1978a,b, 1979, 1981a,b, 1982) let us consider a particular model in which stress concentrations are very highly localized. This model assumes a linear array of fibres, as if the fibres are laid out alongside one another in a single plane, with a load-sharing rule that distributes load to the nearest neighbours of a failed fibre or group of consecutive failed fibres. This is codified by the stress concentration factor $K_r = 1 + r/2$ whose interpretation is that a surviving fibre with r immediately adjacent failed neighbours receives a load K_r times its original load. If there are failed neighbours on both sides then r is the sum of the numbers of failures on each side (Figure 9.7). Since this model essentially arose in the papers of Harow and Phoenix, we call it the Harlow-Phoenix model in the subsequent discussion. The model may be extended to the case where K_r takes some other value than $1 + r/2$ but in that case the interpretation has to be that some load is lost, for instance into the surrounding matrix.

Figure 9.7 An example of stress concentration factors in the simple one-dimensional model.

Of course, this model is itself a simplification in at least two respects, namely the restriction to linear arrays and the nearest-neighbour stress concentrations. One obvious question is: what are the 'true' stress concentrations and how can one attempt to categorize them? Theoretical studies of this question were carried out using the method of shear lag analysis by Hedgepeth (1961), Hedgepeth and Van Dyke (1967), Van Dyke and Hedgepeth (1969) and a series of papers by Fukuda and co-workers, e.g. Fukuda and Kawata (1976), Fukunaga, Chou and Fukuda (1984). These papers largely lead to numerically determined stress concentration factors, since the differential equations which they contain are too hard to solve analytically. Smith *et al.* (1983) proposed some plausible but artificial models for computing stress concentration factors in a three-dimensional array, but these are still oversimplified in that, for instance, they reduce to the Harlow–Phoenix model in the case of a linear array.

Balancing the need for physical realism and mathematical simplicity, we are led to the following model for a local load-sharing system with general geometry. When a group of r adjacent fibres has failed, the stress concentration factor is K_r and the number of unfailed neighbours over which the load is distributed is q_r. A further refinement is to allow the ineffective length δ also to depend on r: when r adjacent fibres have failed the load is distributed along the bundle for a distance δ_r. This model has the disadvantage of ignoring the local geometry (e.g. in a three-dimensional array, no account is taken of whether the r failures occur in a nearly circular group or a long thin ellipse) but the advantage of allowing different degrees of stress concentration to be encompassed in one general model, the Harlow–Phoenix model $K_r = 1 + r/2$, $q_r = 2$ being the most concentrated case. Note that the Harlow–Phoenix model satisfies $(K_r - 1)q_r = r$ and this might be thought a useful relationship in general, reflecting the fact that total load on the system is conserved. However, in general we only require $(K_r - 1)q_r \leq r$ as it is possible that some of the load is absorbed into the matrix or by more distant fibres. The model just described is one that has been adopted in a whole series of papers, e.g. Batdorf (1982), Batdorf and Ghaffarian (1982), Smith *et al.* (1983), Watson and Smith (1985), Bader, Pitkethly and Smith (1987).

9.10* EXACT CALCULATIONS

In this section we show how a natural extension of the argument that led to (9.17) also gives rise to a recursive formula for the distribution of static strength of a bundle of fibres under a very general load-sharing rule. We then go on to consider some more efficient schemes that are available for particular load-sharing rules, in particular the Harlow–Phoenix load-sharing rule described in the previous section.

A general formula.
Consider a bundle of n fibres labelled $1, 2, \ldots, n$ and let $N = \{1, 2, \ldots, n\}$ denote the set of fibres. Assume for each subset S of N that there exist load-sharing coefficients $\lambda_i(S) \geq 0$, $i \in S$, with the interpretation that, if all the fibres outside S fail and if the (original) load per fibre on the system is x, then the load on fibre i is $\lambda_i(S)x$. The assumptions are:

$$\sum_{i \in S} \lambda_i(S) > 0 \quad \text{for all} \quad S \subseteq N, S \neq \emptyset, \tag{9.31}$$

$$\text{If } S \subseteq S' \subseteq N \quad \text{then} \quad \lambda_i(S) \geq \lambda_i(S') \quad \text{for all } i \in S. \tag{9.32}$$

Condition (9.31) is essentially introduced to ensure that the whole of the load does not disappear; it is very much weaker than assuming $\sum_{i \in S} \lambda_i(S) = n$, which would be natural in many contexts though, for the reason pointed out at the very end of section 9.9, is not something we want to assume automatically. Condition (9.32) is an important condition known as **monotonicity** of the load-sharing rule: it states that the load on a surviving fibre is never decreased as a result of additional failures occuring. Apart from being physically intuitive, this property helps to make the failure process mathematically well-defined, since otherwise the failure or survival of the system could depend on the order in which supposedly simultaneous failures are assumed to have occurred.

For $i \in N$, let F_i denote the cumulative distribution function for the strength of fibre i. We assume that individual fibre strengths are independent random variables. For $S \subseteq N$, let $G(x; S)$ denote the probability that a system consisting of just the fibres in S fails under a load per fibre in N of x (i.e. when the total load in S is nx). We have $G(x; \emptyset) = 1$ for all $x > 0$, and $G(x; \{i\}) = F_i(x)$. Then

$$G(x; S) = \sum_{S' \subseteq S, S' \neq \emptyset} (-1)^{|S'|+1} \left\{ \prod_{i \in S'} F_i(x\lambda_i(S)) \right\} G(x; S - S'), x > 0. \quad (9.33)$$

The proof of this is via the inclusion-exclusion formula. For $i \in S$ let A_i denote the event that fibre i has strength at most $x\lambda_i(S)$ and the system S fails under load x. The event whose probability is required is $\cup_{i \in S} A_i$. The proof is completed by applying the inclusion-exclusion formula (9.8) and noting that, by independence of fibre strengths together with (9.32) (which essentially ensures that the failure mechanism does not depend on the order of individual fibre failures) we have

$$\Pr\left\{ \bigcap_{i \in S'} A_i \right\} = \left\{ \prod_{i \in S'} F_i(x\lambda_i(S)) \right\} G(x; S - S').$$

Formula (9.33) was given by Harlow, Smith and Taylor (1983). In the case of equal load-sharing it can be shown after some manipulation that it is equivalent to (9.17), but it is clearly much less efficient computationally since (9.17) involves just a linear sum whereas (9.33) involves a sum over all subsets. However, it is the most general formula available, encompassing virtually any sensible load-sharing rule.

Recursions for linear bundles

An alternative iterative scheme was developed by Harlow and Phoenix (1981a,b, 1982) for the Harlow–Phoenix load-sharing rule, which simplifies the computation by exploiting the linearity of the bundle geometry. For load per fibre x, bundle size n and a fixed $k > 0$, let $A_n^{[k]}(x)$ denote the event that, somewhere in the bundle, there is a group of k consecutive fibres which fail under the action of the load-sharing rule. Let $\bar{A}_n^{[k]}(x)$ denote the complement of $A_n^{[k]}(x)$, and let $G_n^{[k]}(x)$, $Q_n^{[k]}(x) = 1 - G_n^{[k]}(x)$ denote the probabilities of $A_n^{[k]}(x)$ and $\bar{A}_n^{[k]}(x)$ respectively.

Consider the case $k = 2$. For much of the discussion the dependence on x is assumed and therefore need not be indicated explicitly. Suppose the strength of fibre n is X_n, define intervals $I_0 = [0, x)$, $I_1 = [x, K_1 x)$, $I_2 = [K_1 x, K_2 x)$, $I_3 = [K_2 x, \infty)$, and let p_i denote $\Pr\{X_n \in I_i\}$ for $i = 0, 1, 2, 3$. Also define

$$Q_n^{[2]}[i] = \Pr\{\bar{A}_n^{[2]} | X_n \in I_i\}$$

and of course $Q_n^{[2]} = \sum_{i=0}^{3} p_i Q_n^{[2]}[i]$.

These quantities satisfy the initial conditions $Q_0^{[2]} = Q_1^{[2]} = 1$, $Q_2^{[2]} = (1 - p_0)^2 + 2p_0(p_2 + p_3)$ and the recursive relations

$$Q_n^{[2]}[3] = Q_n^{[2]}[2] = Q_{n-1}^{[2]}, \tag{9.34}$$

$$Q_n^{[2]}[1] = \sum_{i=1}^{3} p_i Q_{n-1}^{[2]}[i], \tag{9.35}$$

$$Q_n^{[2]}[0] = p_2 Q_{n-1}^{[2]}[1] + p_3 Q_{n-1}^{[2]}[3] \tag{9.36}$$

Equations (9.34)–(9.36) may be combined to give a matrix recursive equation of the form

$$Q^{[2]} Q_{n-1}^{[2]} = Q_n^{[2]}, n \geq 2 \tag{9.37}$$

where $Q^{[2]}$ is a 3×3 matrix and $Q_n^{[2]}$ a vector with entries $Q_n^{[2]}[i]$ for $i = 0, 1, 3$.

Equation (9.37) may be solved exactly by diagonalizing $Q^{[2]}$, or alternatively an asymptotic solution for $Q_n^{[2]}$ exists of the form

$$Q_n^{[2]}(x) = \{\lambda^{[2]}(x)\}^n \{\pi^{[2]}(x) + o_n^{[2]}(x)\} \tag{9.38}$$

where $o_n^{[2]}(x) \to 0$ as $n \to \infty$ and $\pi^{[2]}(x)$ is a boundary term tending to 1 as $x \to 0$.

This argument may essentially be extended to calculate $Q_n^{[k]}(x)$ for any k except that the matrix equation (9.37) is now an equation on $2^k - 1$ dimensions. This quickly becomes intractable as k increases but it appears that a good approximation to $Q_n^{[n]}(x)$ exists for quite moderate k and results of the form (9.38) exist for any k. This method leads to very good approximations for bundle failure probabilities in a computationally tractable manner. The main disadvantage is that the method is restricted to essentially just the Harlow–Phoenix load-sharing rule, though some extensions (e.g. where the load is transferred to second-nearest neighbours as well as nearest neighbours) may be handled by the same approach.

A more general approach

The approach just described is based on conditioning on the strength category of the last fibre in the bundle. A more general approach would be to condition on certain key configurations of fibres appearing at the end of the bundle. This idea underlies a generalization developed by Kuo and Phoenix (1987). The technical details are too complicated to describe here, but the approach has a number of advantages including more efficient computations, more powerful asymptotic results and an extension from the static strength case to time-to-failure calculations under assumptions similar to those in section 9.7.

Like the Harlow–Phoenix recursion, this exploits the linearity of the bundle and its application is really restricted to linear arrays, though there are some extensions which it can handle. As far as the current 'state of the art' is concerned, only the very inefficient formula (9.33) appears to be available in complete generality, but algorithms such as those of Harlow and Phoenix, and Kuo and Phoenix, lead to much more efficient approximate computations in certain situations. Apart from their interest in themselves, these algorithms have been used to 'calibrate' some more general but less accurate approximations such as the ones to be discussed in the next section.

Equations (9.30) and (9.38), with the extension of the latter to $Q_n^{[k]}$ for general k, are suggestive of a general form of approximation

$$F_{n,L}(x) \approx 1 - \{1 - W(x)\}^{nL} \tag{9.39}$$

in which W is called the **characteristic distribution function** of the system. Although the existence of W has been established rigorously only for the Harlow–Phoenix and Kuo–Phoenix classes of model, and therefore not for general three-dimensional local load-sharing, it is a useful concept in general. The intuitive interpretation of (9.39) is that a system of length L and n fibres behaves like a weakest-link system of independent components, generated by the distribution function W. Many of the cruder approximations to system failure probabilities, including those described in the next section, are of this general structure.

9.11* APPROXIMATIONS FOR LOCAL LOAD-SHARING SYSTEMS

We return to the model described at the end of section 9.9, with q_r, K_r and δ_r denoting the number of overloaded neighbours, the stress concentration factor on those neighbours and the ineffective length associated with a group of r consecutive failed fibres in the system. The model will be assumed to be as depicted in Figure 9.6, a 'chain of bundles' with n fibres of length L, and let $m = L/\delta$ denote the number of bundles in the chain, δ being the initial ineffective length. For an individual fibre of length a we assume the Weibull distribution function

$$F_a(x) = 1 - \exp\{-a(x/x_1)^\eta\}, \, x > 0, \, x_1 > 0, \, \eta > 0. \tag{9.40}$$

Note that for small x, there is the approximation

$$F_a(x) \approx a(x/x_1)^\eta. \tag{9.41}$$

Following terminology introduced by Batdorf (1982), define a 'k-plet' to be a group of k consecutive failed fibre sections and let N_k denote the expected number of k-plets in the system, for a fixed small load x.

Using the approximation (9.41), the expected number of 1-plets is

$$N_1 \approx mn\delta(x/x_1)^\eta.$$

This essentially uses the fact that there are mn short fibre sections of length δ.

A 2-plet is created from a 1-plet (i.e. a single failed fibre of length δ) when one of its neighbours fails. Since a 1-plet is surrounded by q_1 overloaded neighbours receiving load K_1x over a length δ_1, the conditional probability that a 1-plet grows into a 2-plet is approximately $q_1\delta_1(K_1x/x_1)^\eta$. Note that this neglects the possibility that one of the neighbours was failed initially, but the effect of this will be negligible provided $K_1^\eta \gg 1$, a reasonable assumption when η itself is large. The approximation also neglects boundary effects. Based on this approximation we have

$$N_2 \approx mn\delta(x/x_1)^\eta q_1\delta_1(K_1x/x_1)^\eta.$$

Applying the same reasoning successively to r-plets for $r = 3, \ldots, k$ we have

$$N_k \approx nL(x/x_1)^{k\eta} \prod_{r=1}^{k-1} (q_r\delta_r K_r^\eta). \tag{9.42}$$

This provides a very good approximation for small k, but as k increases the approximation becomes much less satisfactory, for several reasons. The simplest of these is that the passage from (9.40) to (9.42) is effectively being applied with x replaced by K_rx for r up to $k-1$, and therefore requires that $K_{k-1}x$ should be small enough for this to be valid. This point (which could, in itself, be easily corrected) is not the only one restricting the viability of (9.42): the whole approach based on summation of probabilities for non-disjoint events only works if those probabilities are small, and this restricts the size of k. A much more rigorous approach to these questions can be taken using extreme value theory (Smith 1980, 1983; Smith *et al.* 1983) but for the present discussion we content ourselves with these rather heuristic points.

The effect of this discussion is that we cannot simply apply (9.42) for very large k, but must instead look for the k that gives the best approximation. Fortunately there is a physical interpretation of this, because as the k-plet grows it will reach a critical size $k = k^*$ after which the stress concentrations on neighbouring fibres become so large that further expansion of the group of failures, to the point where the whole system fails, becomes inevitable.

Applying (9.42) with $k=k^*$, we find

$$N_{k^*} \approx (x/x_L^*)^{\eta^*}$$

where

$$\eta^* = k^*\eta, \; x_L^* = x_1 \left\{ nL \prod_{r=1}^{k-1} (q_r \delta_r K_r^\eta) \right\}^{-1/k^*\eta}.$$

Carrying through the interpretation of k^* as a critical size of failure configuration, we may (to the level of approximation being considered) equate the event of bundle failure with $\{N_{k^*} \geq 1\}$: using a Poisson approximation for the distribution of N_{k^*} then leads us to write the probability of bundle failure as $1 - \exp\{-(x/x_L^*)^{\eta^*}\}$, a Weibull distribution with scale parameter x_L^* and shape parameter η^*.

We have passed over the practical choice of k^*. Based on theoretical calculations and comparison with the Harlow–Phoenix results in the linear case, Smith (1983) suggested choosing as k^* the value of k that minimizes the right-hand side of (9.42). Some indication of the consequences of this will be clear from the example in the next section.

This approximation is only concerned with static strength – in effect, the generalization of the Daniels bundle to the case of local load-sharing. A similar generalization exists for time-dependent failure under the assumptions on single fibres of section 9.7. This was developed by Phoenix and Tierney (1983) following fundamental theoretical work of Tierney (1982).

9.12* STATISTICAL APPLICATIONS OF LOAD-SHARING MODELS

For this final section we return to the statistical theme of the book, describing two statistical applications of the foregoing results. The first is a return to the example of section 4.8, where we focus on the relation between the fitted Weibull distributions for fibres and bundles in the light of the approximations developed in section 9.11. The second example describes a more advanced application in which the probabilisitic theory is used as an inferential technique, to obtain information about the true stress concentration coefficients.

The data in section 4.8 were based on experiments on carbon fibres of diameter approximately 8 μm. (1 μm., also called a micron, is 10^{-6} metres.) In order to apply the theory of section 9.11, we must postulate reasonable values for δ_r, q_r and K_r for small r. Assuming the ineffective length δ to be 5 fibre diameters, we postulate δ to be 0.04 mm. For δ_r, considerations analogous to continuum fracture mechanics suggest that this should grow proportional to $\sqrt{r}$, so we write $\delta_r = \delta\sqrt{r}$.

For the K_r, think of the fibres in cross section as forming a square grid. When one fibre fails, we may assume that the stress on that fibre is spread equally over four neighbours, each receiving a stress concentration factor of $5/4$. When two consecutive fibres fail, load is spread to the six nearest neighbours so the stress concentration factor on each is $8/6$. When three adjacent fibres fail, two configurations are possible: either they are in a straight line or they form a right-angled triangle. Taking the latter of these as more probable we deduce a stress concentration factor of $10/7$ spread over seven neighbours. These are of course simplified calculations but theoretical work has suggested that they are a reasonable approximation to the true stress concentration factors.

Extending this reasoning as far as $r=9$, geometrical reasoning suggests $q_r = 4, 6, 7, 8, 9, 10, 10, 11, 12$ for $r = 1, \ldots, 9$, and $K_r = 1 + r/q_r$. We are now in a position to apply the approximations of section 9.11.

We take $n = 1000$, $\delta = 0.04$, $\delta_r = \delta\sqrt{r}$, $x_1 = 4.775$, $\eta = 5.6$. The last two values are derived from the maximum likelihood fit to single fibres. As an example of the calculations, Table 9.5 gives the calculated value for x_L^* and η^* for each value of k in the case $L = 20$ mm. It is seen that there is very little variability in x_L^* as k varies over the range 5–9, the largest value being when $k = 6$.

Table 9.5 Estimated x_L^* and η^* in terms of k given $L = 20$ mm

k	1	2	3	4	5	6	7	8	9	10
x_L^*	0.8146	2.0141	2.5709	2.8172	2.9218	2.9572	2.9567	2.9295	2.8946	2.8564
η^*	5.6	11.2	16.8	22.4	28.0	33.6	39.2	44.8	50.4	56.0

Table 9.6 gives the resulting theoretical values for $L = 20, 50, 150, 300$, when the k that produces the largest x_L^* is taken, together with the results of separate Weibull fits to the four data sets. Standard errors for the fitted x_L^* are all about 0.03, while those for the fitted η^* are in the range 2–3.

Comparison of theoretical and fitted values shows good agreement in x_L^*, at least for lengths 20 and 50, but poor agreement in η^*. However, in view of the near-agreement of the results over several consecutive values of k, it is clear

Table 9.6 Comparison of theoretical and fitted Weibull parameters for impregnated bundles

Length (mm)	Critical k	Theor. x_L^*	Theor. η^*	Fitted x_L^*	Fitted η^*
20	6	2.96	33.6	2.90	20.3
50	7	2.89	39.2	2.88	19.4
100	7	2.86	39.2	2.76	19.0
200	7	2.81	39.2	2.61	12.9

that the theoretical analysis does not pin down η^* at all precisely. The discrepancies at lengths 100 and 200 are more worrying: although the theoretical analysis is uncertain over the precise choice of k, it certainly predicts an increase of k and hence η^* as L increases, so the fact that the fitted η^* decreases suggests an increase in experimental variability due to other sources of variation. Subsequent experimental results by M. Pitkethly (Bader, Pitkethly and Smith, 1987) gave closer agreement between experiment and theory, but there remains much work to be done on these models.

Our second example (Wolstenholme and Smith, 1989) is based on using load-sharing calculations to construct a probability distribution for the complete point process of fibre failures in a composite. This is then used to construct a likelihood function based on the observed data, and this in turn leads to statistical inference about the parameters of the process, in particular the stress concentration factors.

The experimental data considered here are based on experiments on carbon tows embedded in a glass-epoxy composite (Bader and Pitkethly, 1986). The tows actually consist of 1000 fibres and are essentially the same as the impregnated bundles of the previous example, but for the purpose of the present experiment they are thought of as single fibres, seven of them being embedded in the form of a linear array within the glass-epoxy composite. The seven tows are arranged in parallel with a distance d between tow centres. In practice four values of d have been used: 1.5, 1.0 and 0.5 mm, and the closest possible distance at which the tows are touching (about 0.25 mm.). The whole system is placed under increasing tension, and the complete pattern of breaks in the carbon tows up to a certain tension can be observed. As an example, Figure 9.8 shows five such patterns with separation distance 1.0 mm. From visual inspection of this, there seems to be some correlation between breaks in neighbouring tows, but it is nothing like a perfect correlation. The same figures with $d = 0.5$ and 1.5 mm show, respectively, a much higher degree of correlation (so that most failures spread right across the bundle) and no visible correlation at all.

We can imagine this system as consisting of single tows obeying the usual Weibull assumptions, with an ineffective length δ of the order of 1 mm, and unknown stress concentration factors K_r. It is not clear how the K_rs should be specified but one simple parametric form is $K_r = 1 + \sqrt{r/F}$ where F is an unknown constant. The fact that F is unspecified reflects the absorption of some of the load by the glass-epoxy combination, which in this case does not act purely as a ductile matrix but is capable of bearing load itself.

Again using the chain-of-bundles model, for each short cross-section of length δ we know which tows in that cross-section failed, and also what was the load on the system at the point when failure occurred. From this it is possible to construct a likelihood function for that cross-section, essentially giving the probability of that configuration in terms of the Weibull parameters

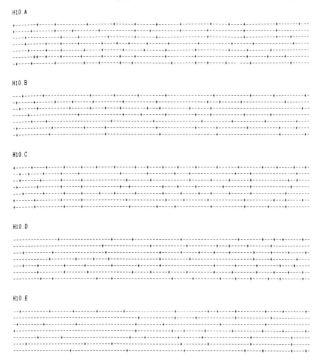

Figure 9.8 The point process of breaks in carbon fibre tows, for 5 test specimens.

on the tows and F. Since there are only seven tows it is possible to do these calculations exactly, though it is computationally intensive.

In view of the computational difficulties of calculating and maximizing a full likelihood function, the main approach adopted was to assume the Weibull parameters from single-tow experiments, and to estimate the one unknown parameter F by an approximate maximum likelihood procedure. For example, with spacing 1.0 mm and an assumed δ of 2 mm, the point estimate of F obtained was 48 with a 95% confidence interval (42, 56) based on the likelihood ratio. An extension allowed for the joint estimation of F and the Weibull scale parameter with only the Weibull shape parameter assumed known in advance – this appears to be a useful extension since there is some evidence that the scale parameter is different for a tow in this system than for a single tow treated on its own.

This very brief description is only an outline of the idea, and the paper (Wolstenholme and Smith, 1989) describes both the technical details and the interpretation of the results in some detail, but it is included here to illustrate the point that the models in sections 9.5–11 are not only for use in probabilistic calculations, but also can be used constructively to analyse and interpret data on the behaviour of complicated systems.

Appendix
The Delta method

In section 3.2 we introduced the result that if $\hat{\theta}$ is the maximum likelihood estimate of θ with approximate standard error a, then $g(\hat{\theta})$, the MLE of $g(\theta)$, has approximate standard error $|g'(\hat{\theta})|a$, where $g'(\hat{\theta})$ means $dg(\theta)/d\theta$ evaluated at $\hat{\theta}$. This appears as (3.4). The generalization to the case where θ is m-dimensional appears as (3.3). Here we shall discuss these results, first in rather general terms and then in the context of maximum likelihood estimation.

Let T_n for $n = 1$, 2, 3, ... be a sequence of random variables such that $n^{1/2}(T_n - \theta)$ has asymptotically, as $n \to \infty$, a Normal distribution with mean 0 and variance $w(\theta)$, where θ is a real parameter. We write this as

$$n^{1/2}(T_n - \theta) \xrightarrow{d} N(0, w(\theta)). \tag{A.1}$$

Then if g is a function with

$$g'(\theta) = \frac{dg(\theta)}{d\theta} \neq 0$$

it follows that

$$n^{1/2}\{g(T_n) - g(\theta)\} \xrightarrow{d} N(0, (g'(\theta))^2 w(\theta)). \tag{A.2}$$

The proof of this result hinges on the Taylor expansion:

$$g(T_n) = g(\theta) + (T_n - \theta)\{g'(\theta) + \varepsilon_n\}, \tag{A.3}$$

where ε_n is an error term. Rearranging (A.3) yields

$$n^{1/2}\{g(T_n) - g(\theta)\} = n^{1/2}(T_n - \theta)\{g'(\theta) + \varepsilon_n\}. \tag{A.4}$$

From (A.4) it is straightforward to show that the error term ε_n is sufficiently small for large enough n so that $n^{1/2}\{g(T_n) - g(\theta)\}$ and $n^{1/2}(T_n - \theta)g'(\theta)$ have the same asymptotic distribution. Hence, we obtain (A.2) as required.

Furthermore, if $g'(\theta)$ and $w(\theta)$ are continuous functions it follows from (A.2) that

$$\frac{n^{1/2}\{g(T_n)-g(\theta)\}}{|g'(T_n)|w^{1/2}(T_n)} \xrightarrow{d} N(0, 1). \tag{A.5}$$

In the context of ML estimation we can take $T_n = \hat{\theta}_n$, the MLE of θ based on a sample of size n. Under mild conditions on the likelihood function (A.1) holds with

$$w(\theta) = \left\{ E\left[-\frac{d^2 \log f(x; \theta)}{d\theta^2} \right] \right\}^{-1}, \tag{A.6}$$

where $f(x; \theta)$ is the density function of an observation. In practice this tells us that $g(\hat{\theta}_n)$ is approximately Normally distributed with mean $g(\theta)$ and standard error $n^{-1/2}|g'(\hat{\theta}_n)|w^{1/2}(\hat{\theta}_n)$. Usually $n^{-1}w(\hat{\theta}_n)$ is replaced by the reciprocal of the sample information, that is by

$$v = \left\{ -\frac{d^2 \log l(\hat{\theta}_n)}{d\theta} \right\}^{-1}, \tag{A.7}$$

where $l(\hat{\theta}_n)$ is the log-likelihood function evaluated at $\theta = \hat{\theta}_n$. Hence (3.4) is obtained.

A generalization to the case in which $\theta = (\theta_1, \theta_2, \ldots, \theta_m)$ is an m-dimensional vector parameter, $T_n = (T_{1n}, T_{2n}, \ldots, T_{mn})$ is an m-dimensional sequence of random vectors and $g_1, g_2, \ldots, g_p$ are p functions of m variables is as follows. Suppose the asymptotic distribution of $n^{1/2}(T_n - \theta)$ is m-variate Normal with mean vector 0 and covariance matrix $W(\theta) \neq 0$ with entries w_{jk}. Then by a similar argument to the above it can be shown that the joint asymptotic distribution of

$$n^{1/2}\{g_j(T_{1n}, \ldots, T_{mn}) - g_j(\theta_1, \ldots, \theta_m)\}; j = 1, 2, \ldots, p$$

is p-variate Normal with mean vector 0 and covariance matrix GWG^T, where the (j, k)th entry of G is $\partial g_j/\partial \theta_k$. Putting this in terms of MLEs $\hat{\theta}_n = (\hat{\theta}_{1n}, \ldots, \hat{\theta}_{mn})$, it follows that in large samples the joint distribution of $g_j(\hat{\theta}_{1n}, \ldots, \hat{\theta}_{mn})$ for $j = 1, 2, \ldots, p$ is approximately p-variate Normal with mean vector $(g_1(\theta), \ldots, g_p(\theta))$ and estimated covariance matrix GV_nG^T, where G is evaluated at $\theta = \hat{\theta}_n$ and V_n is the inverse of the observed information matrix of the sample. In the special case $p = 1$, we have that $g(\hat{\theta}_n)$ is approximately Normal with mean $g(\theta)$ and standard error (3.3).

Example

Let $\hat{S}_n(t)$ be the empirical survivor function of a sample of size n with no censored observations; see (2.13). In section 2.9 we investigated the properties of $\hat{S}_n(t^*)$ and $\hat{H}_n(t^*) = -\log \hat{S}_n(t^*)$, where t^* is fixed. In particular, using standard results for the Binomial distribution we obtained the standard error for $\hat{S}_n(t^*)$, (2.14), as

$$\left\{ \frac{\hat{S}_n(t^*)[1 - \hat{S}_n(t^*)]}{n} \right\}^{1/2}. \tag{A.8}$$

By the Central Limit Theorem $\hat{S}_n(t^*)$ satisfies (A.1) with $T_n = \hat{S}_n(t^*)$, $\theta = S(t^*)$ and $w(\theta) = \theta(1 - \theta)$. If we take $g(\theta) = -\log \theta$, (A.2) and continuity gives (A.5) with $g'(T_n) = -1/T_n$. Hence the standard error of $\hat{H}_n(t^*) = -\log \hat{S}_n(t^*)$ is $n^{-1/2}|-1/T_n|\{T_n(1 - T_n)\}^{1/2}$. That is, $n^{-1/2}\{[1 - \hat{S}_n(t^*)]/\hat{S}_n(t^*)\}^{1/2}$, which is precisely the result given in (2.16).

References

Abramowitz, M. and Stegun, I. A. (eds) (1972) *Handbook of Mathematical Functions*. Dover, New York.

Agrawal, A. and Barlow, R. E. (1984) A survey of network reliability and domination theory. *Operations Research*, **32**, 478–92.

Agrawal, A. and Satyanarayna, A. (1984) An O(|E|) time algorithm for computing the reliability of a class of directed networks. *Operations Research*, **32**, 493–515.

Aitkin, M. and Clayton, D. (1980) The fitting of exponential, Weibull and extreme value distributions to complex censored survival data using GLIM. *Applied Statistics*, **29**, 156–63.

Akman, V. E. and Raftery, A. E. (1986) Asymptotic inference for a change-point Poisson process. *Ann. Statist.*, **14**, 1583–90.

Andersen, P.K. (1982) Testing the goodness-of-fit of Cox's regression and life model. *Biometrics*, **38**, 67–72.

Ansell, J. I. (1987) Analysis of lifetime data with covariates in *Proc. 6th Natl Reliability Conference, Birmingham*. Warrington, National Centre of Systems Reliability.

Ansell, R. O. and Ansell, J.I. (1987) Modelling the reliability of sodium sulphur cells. *Reliabil. Engng.*, **17**, 127–37.

Ansell, J. I. and Phillips, M. J. (1989) Practical problems in the analysis of reliability data (with discussion). *Appl. Statist.* **38**, 205–47.

Ascher, H. (1982) Regression analysis of repairable systems reliability in *Proc. NATO Advanced Study Institute on Systems Effectiveness and Life Cycle Costings*. Springer, New York.

Ascher, H. and Feingold, H. (1984) *Repairable Systems Reliability*. Marcel Dekker, New York.

Atkinson, A. C. (1970) A method of discriminating between models. *J. R. Statist. Soc. B*, **32**, 323–45.

Atkinson, A.C. (1985) *Plots, Transformations and Regression*. Oxford University Press, London.

Bader, M. G. and Pitkethly, M. J. (1986) Probabilistic aspects of the strength and modes of failure of hybrid fibre composites in *Mechanical Characterisation of Fibre Composite Materials*, (ed. R. Pyrz.) Aalborg University, Denmark.

Bader, M. G., Pitkethly, M. J. and Smith, R. L. (1987) Probabilistic models for hybrid composites in *Proceedings of ICCM-VI and ECCM-2*, **5**, ed. F.L. Matthews, N. C. R. Buskell, J. M. Hodgkinson and J. Morton, Elsevier Applied Science, London.

Barbour, A. D. (1975) A note on the maximum size of a closed epidemic. *J. R. Statist. Soc. B*, **37**, 459–60.

Barbour, A. D. (1981) Brownian motion and a sharply curved boundary. *Adv. Appl. Prob.*, **13**, 736–50.

Barlow, R. E. and Proschan, F. (1981) *Statistical Theory of Reliability and Life Testing: Probability Models*. To Begin With, Silver Spring, Maryland. (Second edition; first edition published by Holt, Reinhart and Winston 1975.)

Barnett, V. (1982) *Comparative Statistical Inference*. Wiley, New York.

Batdorf, S. B. (1982) Tensile strength of unidirectionally reinforced composites I. *J. Reinf. Plast. Compos.*, **1**, 153–64.

Batdorf, S. E. and Ghaffarian, R. (1982) Tensile strength of unidirectionally reinforced composites II. *J. Reinf. Plast. Compos.*, **1**, 165–76.

Beasley, J.D. and Springer, S. G. (1977) Algorithm 111, *Appl. Statist.*, **26** 118–21.

Bennett, S. (1983) Analysis of survival data by the proportional odds model. *Statistics in Medicine*, **2**, 273–77.

Berger, J. O. (1985) *Statistical Decision Theory and Bayesian Analysis.* Springer-Verlag, Berlin.

Bergman, B. (1985) On reliability theory and its application. *Scand. J. Statist.*, **12**, 1–41.

Borges, W. de S. (1983) On the limiting distribution of the failure time of fibrous materials. *Adv. Appl. Prob.*, **15**, 331–48.

Box, G. E. P. and Cox, D. R. (1964) An analysis of transformations (with discussion). *J. R. Statist. Soc. B*, **26**, 211–52.

Brindley, E. C. and Thompson, W. A. (1972) Dependence and ageing aspects of multivariate survival. *J. Amer. Statist. Assoc.*, **67**, 822–30.

Buehler, R. J. (1957) Confidence limits for the product of two binomial parameters. *J. Amer. Statist. Assoc.*, **52**, 482–93.

Buffon (1775) *Histoire naturelle generale et particuliere.* Imprimerie Royale Paris.

Burr, I. W. (1942) Cumulative frequency distributions. *Ann. Math. Statist.*, **13**, 215–32.

Chambers, J. M., Cleveland, W. S., Kleiner, B., and Tukey, P. A. (1983) *Graphical Methods for Data Analysis.* Wadsworth International, Belmont CA.

Chao, A. and Huwang, L.-C. (1987) A modified Monte Carlo technique for confidence limits of system reliability using pass-fail data. *IEEE Trans. Reliability*, **R-36**, 109–12.

Chatfield, C. (1980) *The Analysis of Time Series.* Chapman and Hall, London.

Chatfield, C. (1983) *Statistics for Technology* 3rd edn, Chapman and Hall, London.

Cheng, R. C. H. and Amin, N. A. K. (1983) Estimating parameters in continuous univariate distributions with a shifted origin. *J. R. Statist. Soc. B*, **45**, 394–403.

Chhikara, R. S. and Folks, J. L. (1977) The inverse Gaussian distribution as a lifetime model. *Technometrics*, **19**, 461–8.

Clayton, D. G. (1978) A model for association in bivariate life tables and its application in epidemiological studies of fasilial tendency in chronic disease incidence. *Biometrika* **65**, 141–51.

Coleman, B. D. (1956) Time dependence of mechanical breakdown phenomena. *J. Apppl. Phys.*, **27**, 862–6.

Coles, S. C. (1989) On goodness-of-fit tests for the two parameter Weibull distribution derived from the stabilized probability plot. *Biometrika*, **76**, 593–8.

Cornell, C. A. (1972) Bayesian statistical decision theory and reliability-based design. *Int. conf. on Structural Safety and Reliability, Washington, 1969.* Pergamon Press.

Cox, D. R. (1959) The analysis of exponentially distributed lifetimes with two types of failure. *J. R. Statist. Soc. B*, **21**, 411–21.

Cox, D. R. (1961) Tests of separate families of hypotheses. *Fourth Berkeley Symposium*, 105–23.

Cox, D. R. (1962a) *Renewal Theory.* Methuen, London.

Cox, D. R. (1962b) Further results on tests of separate families of hypotheses. *J. Roy. Statist. Soc. B*, **24**, 406–24.

Cox, D. R. (1972) Regression models and life tables (with discussion). *J. R. Statist. Soc. B*, **34**, 187–202.

Cox, D. R. (1975) Partial likelihood. *Biometrika*, **62**, 269–76.

Cox, D. R. and Hinkley, D. V. (1974) *Theoretical Statistics.* Chapman and Hall, London.

Cox, D. R. and Isham, V. (1980) *Point Processes*. Chapman and Hall, London.

Cox, D. R. and Lewis, P. A. W. (1966) *The Statistical Analysis of Series of Events*. Methuen, London.

Cox, D. R. and Oakes, D. (1984) *Analysis of Survival Data*. Chapman and Hall, London.

Cox, D. R. and Snell, E. J. (1968) A general definition of residuals (with discussion). *J. R. Statist. Soc. B*, **30**, 248–75.

Cox, D. R. and Snell, E. J. (1989) *The Analysis of Binary Data*, 2nd edn. Chapman and Hall, London.

Crawford, D. E. (1970) Analysis of incomplete life test data on motorettes. *Insulation/Circuits*, **16**, 43–8.

Crow, L. H. (1974) Reliability analysis for complex repairable systems in *Reliability and Biometry* (eds F. Proschan and R. J. Serfling). SIAM, Philadelphia.

Crowder, M. J. (1985) A distributional model for repeated failure time measurements. *J. R. Statist. Soc. B*, **47**, 447–52.

Crowder, M. J. (1989) A multivariate distribution with Weibull connections. *J. R. Statist. Soc. B*, **51**, 93–107.

Crowder, M. J. (1990) On some nonregular tests for a modified Weibull model. *Biometrika*, **77**, 499–506.

Crowder, M. J. and Hand, D. J. (1990) *Analysis of Repeated Measures*. Chapman and Hall, London.

D'Agostino, R. B. and Stephens, M. A. (1986) *Goodness-of-fit Techniques*. Marcel Dekker, New York.

Daniels, H. E. (1945) The statistical theory of the strength of bundles of threads. *I. Proc. R. Soc. Lond. A*, **183**, 404–35.

Daniels, H. E. (1974a) An approximation technique for a curved boundary problem. *Adv. Appl. Prob.*, **6**, 194–6.

Daniels, H. E. (1974b) The maximum size of a closed epidemic. *Adv. Appl. Prob.*, **6**, 607–21.

Daniels, H. E. (1989) The maximum of a Gaussian process whose mean path has a maximum, with an application to the strength of bundles of fibres. *Adv. Appl. Probab.*, **21**, 315–33.

Daniels, H. E. and Skyrme, T. H. R. (1985) The maximum of a random walk whose mean path has a maximum. *Adv. Appl. Prob.*, **17**, 85–99.

David, H. A. (1981) *Order Statistics*. Wiley, New York.

David, H. T. (1987) Bayes extraction of component failure information from Boolean module test data. *Commun. Statist.*, **16**(10), 2873–83.

Davis, D. J. (1952) An analysis of some failure data. *J. Amer. Statist. Assoc.*, **47**, 113–50.

DeGroot, M. H. (1970) *Optimal Statistical Decisions*. McGraw-Hill, New York.

DiCiccio, T. J. (1987) Approximate inference for the generalized gamma distribution. *Technometrics*, **29**, 33–40.

Duane, J. T. (1964) Learning curve approach to reliability monitoring, *IEEE t rans*, A-2, 563–6.

Drury, M., Walker, E. V., Wightman, D. W. and Bendell, A. (1987) Proportional hazards modelling in the analysis of computer systems reliability in *Proc. 6th Natl Reliability Conference, Birmingham*. Warrington, National Centre of Systems Reliability.

Dumonceaux, R. and Antle, C E. (1973) Discrimination between the lognormal and Weibull distributions. *Technometrics*, **15**, 923–6.

Dumonceaux, R., Antle, C. E. and Haas, G. (1973) Likelihood ratio test for discrimination between two models with unknown location and scale parameters (with discussion). *Technometrics*, **15**, 19–31.

Easterling, R. G. (1972) Approximate confidence intervals for system reliability. *J. Amer. Statist. Assoc.*, **67**, 220–2.

Eastham, J. F., LaRiccia, V. N. and Schuenemeyer, J. H. (1987) Small sample properties of maximum likelihood estimators for an alternative parametrization of the three-parameter lognormal distribution. *Commun. Statist.*, **16**, 871–84.

Efron, B. (1979) Bootstrap methods: another look at the jackknife. *Ann. Statist.*, **7**, 1–26.

Efron, B. (1982) *The Jackknife, the Bootstrap and Other Resampling Plans*. SIAM, Philadelphia.

Eggwertz, S. and Lind, N. C. (eds) (1985) *Probabilistic Methods in the Mechanics of Solids and Structures*. (IUTAM Symposium, Stockholm 1984). Springer-Verlag, Berlin.

Evans, R. A. (1989) Bayes is for the birds. *IEEE Trans. Reliability*, **R-38**, 401.

Everitt, B. S. and Hand, D. J. (1981) *Finite Mixture Distributions*. Chapman and Hall, London.

Feller, W. (1968) *An Introduction to Probability Theory and Its Applications*, **I**. 3rd Edn, John Wiley, New York.

Freund, J. E. (1961) A bivariate extension of the exponential distribution. *J. Amer. Statist. Assoc.*, **56**, 971–7.

Fukuda, H. and Kawata, K. (1976) On the stress concentration factor in fibrous composites. *Fibre Sci. Technol.*, **9**, 189–203.

Fukunaga, H., Chou, T.-W. and Fukuda, H. (1984) Strength of intermingled hybrid composites. *J. Reinf. Plast. Compos.*, **3**, 145–60.

Funnemark, E. and Natvig, B. (1985) Bounds for the availabilities in a fixed time interval for multistate monotone systems. *Adv. Appl. Prob.*, **17**, 638–55.

Gail, M. (1975) A review and critique of some models in competing risks analysis. *Biometrics*, **31**, 209–22.

Galambos, J. (1978) *The Asymptotic Theory of Extreme Order Statistics*. Wiley, New York.

Gaver, D. P. and Acar, M. (1979) Analytical hazard representations for use in reliability, mortality and simulation studies. *Commun. Statist.*, **8**, 91–111.

Gerstle, F. P. and Kunz, S. C. (1983) Prediction of long-term failure in Kevlar 49 composites in *Long-term Behavior of Composites*, ASTM STP 813, (T. K. O'Brien ed), 263–92, Philadelphia.

Gill, R. and Schumacher, M. (1987) A simple test of proportional hazards assumption. *Biometrika*, **74**, 289–300.

Goel, A. L., (1985) Software reliability models: assumptions, limitations and applicability. *IEEE Trans. Software Engng.*, **SE-11**, 1411–23.

Goel, P. K. (1988) Software for Bayesian analysis: current status and additional needs. *Bayesian Statistics 3*, Proceedings of the 3rd Valencia International Meeting, 173–88, Oxford Univ. Press.

Green, P. J. (1984) Iteratively reweighted least squares for maximum likelihood estimation and some robust alternatives (with discussion). *J. R. Statist. Soc. B* **46**, 149–92.

Greene, A. E. and Bourne, A. J. (1972) *Reliability Technology*. Wiley Interscience, London.

Greenwood, M. (1926) The natural duration of cancer. *Reports on Public Health and Medical Subjects*, **33**, 1–26. HMSO, London.

Griffiths, P. and Hill, I. D. (eds) (1985) *Applied Statistics Algorithms*. Ellis Horwood, Chichester.

Groeneboom, P. (1989) Brownian motion with a parabolic drift and Airy functions. *Prob. Th. Rel. Fields*, **81**, 79–109.

Gumbel, E. J. (1960) Bivariate exponential distribution. *J. Amer. Statist. Assoc.*, **55**, 698–707.

Hagstrom, J. N. and Mak, K.-T. (1987) System reliability analysis in the presence of dependent component failures. *Prob. Eng. Inf. Sci.*, **1**, 425–40.

Harlow, D. G. and Phoenix, S. L. (1978a) The chain-of-bundles models for the strength of fibrous materials I: analysis and conjectures. *J. Composite Materials*, **12**, 195–214.

Harlow, D. G. and Phoenix, S. L. (1978b) The chain-of-bundles models for the strength of fibrous materials II: a numerical study of convergence. *J. Composite Materials*, **12**, 314–34.

Harlow, D. G. and Phoenix, S. L. (1979) Bounds on the probability of failure of composite materials. *Internat. J. Fracture*, **15**, 321–36.

Harlow, D. G. and Phoenix, S. L. (1981a) Probability distributions for the strength of composite materials I: Two-level bounds. *Internat. J. Fracture*, **17**, 347–72.

Harlow, D. G. and Phoenix, S. L. (1981b) Probability distributions for the strength of composite materials II: A convergent sequence of tight bounds. *Internat. J. Fracture*, **17**, 601–30.

Harlow, D. G. Phoenix, S. L. (1982) Probability distributions for the strength of fibrous materials I: Two level failure and edge effects. *Adv. Appl. Prob.*, **14**, 68–94.

Harlow, D. G., Smith, R. L. and Taylor, H. M. (1983) Lower tail analysis of the distribution of the strength of load-sharing systems. *J. Appl. Prob.*, **20**, 358–67.

Hedgepeth J. M. (1961) Stress concentrations in filamentary composite materials. *Technical Note D-882*, NASA, Washington.

Hedgepeth, J. M. and Van Dyke, P. (1967) Local stress concentrations in imperfect filamentary composite materials. *J. Compos. Mater.*, **1**, 294–309.

Hosking, J. R. M. (1985) Algorithm AS215: Maximum likelihood estimation of the parameters of the generalized extreme-value distribution. *Appl. Statist.*, **34**, 301–10.

Hougaard, P. (1984) Life table methods for heterogeneous populations: distributions describing the heterogeneity. *Biometrika*, **71**, 75–83.

Hougaard, P. (1986a) Survival models for heterogeneous populations derived from stable distribution. *Biometrika*, **73**, 387–96.

Hougaard, P. (1986b) A class of multivariate failure-time distributions. *Biometrika*, **73**, 671–8.

Isaacs, G. L., Christ, D. E., Novick, M. R. and Jackson, P. H. (1974) *Tables for Bayesian Statisticians*. Iowa Univ. Press, Ames, IO.

Jelinski, Z. and Moranda, P. B. (1972) Software reliability research in *Statistical Computer Performance Evaluation* (ed. W. Freiberger). Academic Press, London.

Johnson, N. L. and Kotz, S. (1970) *Distributions in Statistics: Continuous univariate distributions I*. Wiley, New York.

Johnson, N. L. and Kotz, S. (1972) *Distributions in Statistics: Continuous multivariate distributions*. Wiley, New York.

Johnson, N. L. and Kotz, S. (1975) A vector multivariate hazard rate. *J. Multi. Analysis*, **5**, 53–5.

Kalbfleisch, J. and Prentice, R. L. (1973) Marginal likelihoods based on Cox's regression and life model. *Biometrika*, **60**, 267–78.

Kalbfleisch, J. and Prentice, R. L. (1980) *The Statistical Analysis of Failure Time Data*. Wiley, New York.

Kaplan, E. L. and Meier, P. (1958) Nonparametric estimation from incomplete observations. *J. Amer. Statist. Assoc.*, **53**, 457–81.

Kaplan, S. (1990) Bayes is for eagles. *IEEE Trans. Reliability*, **R-39**, 130–1.

Kappenman, R. F. (1985) Estimation for the three-parameter Weibull, lognormal and gamma distributions. *Computational Statistics and Data Analysis*, **3**, 11–23.

Kass, R. E., Tierney, L. and Kadane, J. B. (1988) Asymptotics in Bayesian computation. *Bayesian Statistics 3*, Proceedings of the 3rd Valencia International Meeting, 791–9, Oxford Univ. Press.

Kelly, A. and McCartney, L. N. (1981) Failure by stress corrosion of bundles of fibres. *Proc. R. Soc. Lond.* A, **374**, 475–89.

Kennedy, J. B. and Neville, A. M. (1986) *Basic Statistical Methods for Engineers and Scientists*, Harper and Row, New York.

Kimber, A. C. (1985). Tests for the exponential, Weibull and Gumbel distributions based on the stabilized probability plot. *Biometrika*, **72**, 661–3.

Kimber, A. C. (1990) Exploratory data analysis for possibly censored data from skewed distributions. *Appl. Statist.*, **39**, 21–30.

Kozin, F. and Bogdanoff, J. L. (1981) A critical analysis of some probabilistic models of fatigue crack growth. *Eng. Frac. Mech.*, **14**, 59–81.

Kuo, C. C. and Phoenix, S. L. (1987) Recursions and limit theorems for the strength and lifetime distributions of a fibrous composite. *J. Appl. Prob.*, **24**, 137–59.

Langberg, N. and Singpurwalla, N. D. (1985) Unification of some software reliability models via the Bayesian approach. *SIAM J. Sci. Statist. Comput.*, **6**, 781–90.

LaRiccia, V. N. and Kindermann, R. P. (1983) An asymptotically efficient closed form estimator for the three-parameter lognormal distribution. *Commun. Statist.*, **12**, 243–61.

Lau, C-L. (1980) Algorithm 147, *Appl. Statist.*, **29**, 113–4.

Lawless, J. F. (1980) Inference in the generalized gamma and log gamma distributions. *Technometrics*, **22**, 409–19.

Lawless, J. F. (1982) *Statistical Models and Methods for Lifetime Data*. Wiley, New York.

Lawless, J. F. (1983) Statistical methods in reliability (with discussion). *Technometrics*, **25**, 305–35.

Lawless, J. F. (1986) A note on lifetime regression models. *Biometrika*, **73**, 509–12.

Lawless, J. F. (1987) Regression methods Poisson process data. *J. Amer. Statist. Assoc.*, **82**, 808–15.

Lee, L. (1980) Testing adequacy of the Weibull and log linear rate models for a Poisson process. *Technometrics*, **22**, 195–200.

Lee, P. M. (1989) *Bayesian Statistics: An Introduction*. Oxford Univ. Press, New York.

Lehmann, E. L. (1966) Some concepts of dependence. *Ann. Math. Stat.*, **37**, 1137–53.

Lewis, P. A. W. and Shedler, G. S. (1976) Statistical analysis of nonstationary series of events in a data base system. *IBM J. Res. Develop.*, 465–82.

Lieblein, J. and Zelen, M. (1956) Statistical investigation of the fatigue life of deep groove ball bearings. *J. Res. Nat. Bur. Stand.*, **57**, 273–316.

Lindgren, G. and Rootzén, H. (1987) Extreme values: theory and technical applications. *Scand. J. Statist.*, **14**, 241–79.

Lindley, D. V. (1965) *Introduction to Probability and Statistics from a Bayesian Viewpoint II*. Cambridge Univ. Press.

Lindley, D. V. and Scott, W. F. (1984) *New Cambridge Elementary Statistical Tables*. Cambridge Univ. Press.

Lindley, D. V. and Singpurwalla, N. D. (1986a) Reliability and fault tree analysis using expert opinions. *J. Amer. Statist. Assoc.*, **81**, 87–90.

Lindley, D. V. and Singpurwalla, N. D. (1986b) Multivariate distributions for the life lengths of components of a system sharing a common environment. *J. Appl. Prob.*, **23**, 418–31.

Lindley, D. V. and Smith, A. F. M. (1972) Bayes estimates for the linear model (with discussion). *J. R. Statist. Soc.*, B, **34**, 1–41.

Lipow, M. and Riley, J. (1960) *Tables of Upper Confidence Limits on Failure Probability of 1, 2 and 3 Component Serial Systems, Vols. 1 and 2*. U.S. Dept. of Commerce AD-609-100 and AD-636-718, Clearinghouse.

Littlewood, B. and Verrall, J. L. (1981) Likelihood function of a debugging model for computer software reliability. *IEE Trans. Reliabil.*, **R-30**, 145–8.

Lloyd, D. K. and Lipow, M. (1962) *Reliability: Management, Methods and Mathematics.* Prentice Hall, Englewood Cliffs, N. J.

MacLean, C. J. (1974) Estimation and testing of an exponential polynomial rate function within the non-stationary Poisson process. *Biometrika*, **61**, 81–5.

Macleod, A. J. (1989) Algorithm AS R76, A remark on algorithm AS 215: Maximum likelihood estimation of the parameters of the generalized extreme value distribution. *Appl. Statist.*, **38**, 198–9.

Madansky, A. (1965) Approximate confidence limits for the reliability of series and parallel systems. *Technometrics*, **7**, 495–503.

Mann, N. R. and Fertig, K. W. (1973) Tables for obtaining confidence bounds and tolerance bounds based on best linear invariant estimates of parameters of the extreme value distribution. *Technometrics*, **15**, 87–101.

Mann, N. R. and Grubbs, F. E. (1974) Approximately optimal confidence bounds for system reliability based on component data. *Technometrics*, **16**, 335–46.

Mann, N. R., Schafer, R. E. and Singpurwalla, N. D. (1974) *Methods for Statistical Analysis of Reliability and Lifetime Data.* Wiley, New York.

Marshall, A. W. and Olkin, I. (1966) A multivariate exponential distribution. *J. Amer. Statist. Assoc.*, **62**, 30–44.

Martz, H. F. and Duran, B. S. (1985) A comparison of three methods for calculating lower confidence limits on system reliability using binomial component data. *IEEE Trans. Reliability*, **R-34**, 113–20.

Martz. H. F. and Waller, R. A. (1982) *Bayesian Reliability Analysis.* Wiley, New York.

Mastran, D. V. and Singpurwalla, W. D. (1978) A Bayesian estimation of the reliability of coherent structures. *Operations Research*, **20**, 663–72.

McCartney, L. N. and Smith, R. L. (1983) Statistical theory of the strength of fiber bundles. *J. Appl. Mech.*, **50**, 601–8.

Mendenhall, W. and Hader, R. J. (1958). Estimation of parameters of mixed exponentially distributed failure time distributions from censored life test data. *Biometrika*, **45**, 504–20.

Michael, J. R. (1983) The stabilized probability plot. *Biometrika*, **70**, 11–17.

MIL-HDBK-189 (1981) Reliability Growth Management. HQ, *US Army Communications Research and Development Command.* ATTN: DRDCOPT, Fort Monmouth, New Jersey.

Morgan, B. J. T. (1984) *Elements of Simulation.* Chapman and Hall, London.

Moreau, T., O'Quigley, J. and Mesbah, M. (1985) A global goodness-of-fit statistic for the proportional hazards model. *Appl. Statist.*, **34**, 212–8.

Morrison, D. F. (1976) *Multivariate Statistical Methods.* 2nd edn, McGraw-Hill, London.

Musa, J. D. (1975) A theory of software reliability and its applications. *IEEE Trans. Software Engng.*, **SE-1**, 312–27.

Nash, J. C. (1979) *Compact Numerical Algorithms for Computers.* Adam Hilger, Bristol.

Natvig, B. (1985a) Multistate coherent systems in *Encyclopedia of Statistical Sciences*, **5**, (ed. N. L. Johnson and S. Kotz), Wiley, New York, 732–5.

Natvig, B. (1985b) Recent developments in multistate reliability theory in *Probabilistic models in the mechanics of solids and structures*, (ed. S. Eggwertz and N. C. Lind) Springer-Verlag, Berlin, 385–93.

Natvig, B. (1986) Improved bounds for the availabilities in a fixed time interval for multistate monotone systems. *Adv. Appl. Prob.*, **18**, 577–9.

Natvig, B., Sormo, S., Holen, A. T. and Hogasen, G. (1986), Multistate reliability theory - a case study. *Adv. Appl. Prob.*, **18**, 921–32.

Naylor, J. C. and Smith, A. F. M. (1982) Applications of a method for the efficient computation of posterior distributions. *Appl. Statist.*, **31**, 214–25.

Nelson, W. B. (1972) Theory and applications of hazard plotting for censored failure data. *Technometrics*, **14**, 945–65.

Nelson, W. B. (1982) *Applied Life Data Analysis*. Wiley, New York.

Nelson, W. B. and Hahn, G. J. (1972) Linear estimation of a regression relationship from censored data. Part I: simple methods and their application. *Technometrics*, **14**, 247–69.

Oakes, D. (1982) A concordance test for independence in the presence of censoring. *Biometrics*, **38**, 451–5.

O'Connor, P. (1985) *Practical Reliability Engineering*, 2nd edn, Wiley Interscience, Chichester.

Peirce, F. T. (1926) Tensile strength for cotton yarns V – 'The weakest link' – Theorems on the strength of long and of composite specimens. *J. Text. Inst. (Trans.)*, **17**, 355–68.

Pettitt, A. N. and Bin Daud, I. (1990) Investigating time dependence in Cox's proportional hazards model. *Appl. Statist.*, **39**, 313–29.

Phoenix, S. L. (1978) The asymptotic time to failure of a mechanical system of parallel members. *SIAM J. Applied Math.*, **34**, 227–46.

Phoenix, S. L. (1979) The asymptotic distribution for the time to failure of a fiber bundle. *Adv. Appl. Prob.*, **11**, 153–87.

Phoenix, S. L. and Taylor, H. M. (1973) The asymptotic strength distribution of a general fiber bundle. *Adv. Appl. Prob.*, **5**, 200–16.

Phoenix, S. L. and Tierney, L.-J. (1983) A statistical model for the time dependent failure of unidirectional composite materials under local elastic load-sharing among fibers. *Engineering Fracture Mechanics*, **18**, 479–96.

Pike, M. C. and Hill, I. D. (1966) Algorithm 291. *Comm. ACM*, **9**, 684.

Prentice, R. L., Kalbfleisch, J. D., Peterson, A. V., Flournoy, N., Farewell, V. T. and Breslow, N. E. (1978) The analysis of failure times in the presence of competing risks. *Biometrics*, **34**, 541–54.

Prescott, P. and Walden, A. T. (1983) Maximum likelihood estimation of the parameters of the three-parameter generalized extreme-value distribution from censored samples, *J. Statist. Comput. Simul.*, **16**, 241–50.

Press, W. H., Flannery, B. P., Teukolsky, S. A. and Vetterling, W. T. (1986) *Numerical Recipes*. C. U. P., Cambridge.

Proschan, F. (1963) Theoretical explanation of observed decreasing failure rate. *Technometrics*, **5**, 375–83.

Proschan, F. and Sullo, P. (1976) Estimating the parameters of a multivariate exponential distribution *J. Amer. Statist. Assoc.*, **71**, 465–72.

Provan, J. S. and Ball, M. O. (1984) Computing network reliability in time polynomial in the number of cuts. *Operations Research*, **32**, 516–26.

Raftery, A. E. (1988) Analysis of a simple debugging model. *Appl. Statist.*, **37**, 12–22.

Raftery, A. E. and Akman, V. E. (1986) Bayesian analysis for a Poisson process with a change-point. *Biometrika*, **73**, 85–9.

Rice, R. E. and Moore, A. H. (1983) A Monte Carlo technique for estimating lower confidence limits on system reliability using pass-fail data. *IEEE Trans. Reliability*, **R-32**, 336–9.

Rosen, B. W. (1964) Tensile failure of fibrous composites. *AIAA J.*, **2.**, 1985–91.

Rosenblatt, J. R. (1963) Confidence limits for the reliability of complex systems in *Statistical Theory of Reliability*, (ed. M. Zelen), University of Wisconsin Press.

Rosenblatt, M. (1952) Remarks on a multivariate transformation. *Ann. Math. Statist.* **23**, 470–2.

Rossi, F., Fiorentino, M. and Versace, P. (1984) Two-component extreme value distribution for flood frequency analysis. *Wat. Resour. Res.*, **20**, 847–56.

Schoenfeld, D. (1980) Chi-squared goodness-of-fit test for the proportional hazards regression model. *Biometrika*, **67**, 145–53.

Schoenfeld, D. (1982) Partial residuals for the proportional hazards regression model. *Biometrika*, **69**, 239–41.

Scop, P. M. and Argon, A. S. (1969) Statistical theory of strength of laminated composites II. *J. Compos. Mater.*, **3**, 30–47.

Self, S. G. and Liang, K. Y. (1987) Asymptotic properties of maximum likelihood estimators and likelihood ratio tests under nonstandard conditions. *J. Amer. Statist. Assoc.*, **82**, 605–10.

Shaked, M. (1982) A general theory of some positive dependence notions. *J. Mult. An.*, **12**, 199–218.

Shier, D. R. and Lawrence, K. D. (1984) A comparison of robust regression techniques for the estimation of Weibull parameters. *Comm. Statist.*, **13**, 743–50.

Singpurwalla, N. D. (1988) Foundational issues in reliability and risk analysis. *SIAM Review*, **30**, 264–82.

Siswadi and Quesenberry, C. P. (1982) Selecting among Weibull, lognormal and gamma distributions using complete and censored samples. *Naval Research Logistics Quarterly*, **29**, 57–569.

Smith, J. Q. (1988) *Decision Analysis: A Bayesian Approach*, Chapman and Hall, London.

Smith, R. L. (1980) A probability model for fibrous composites with local load-sharing. *Proc. R. Soc. Lond. A*, **372**, 539–53.

Smith, R. L. (1982) The asymptotic distribution of the strength of a series-parallel system with equal load-sharing. *Ann. Probab.*, **10**, 137–71.

Smith, R. L. (1983) Limit theorems and approximations for the reliability of load-sharing systems. *Adv. Appl. Prob.*, **15**, 304–30.

Smith, R. L. (1985) Maximum likelihood estimation in a class of nonregular cases. *Biometrika*, **72**, 67–92.

Smith, R. L. (1989) Nonregular regression. *Submitted for publication*.

Smith, R. L. and Naylor, J. C. (1987) A comparison of maximum likelihood and Bayesian estimators for the three-parameter Weibull distribution. *Appl. Statist.* **36**, 358–69.

Smith, R. L., Phoenix, S. L., Greenfield, M. R., Henstenburg, R. and Pitt, R.E. (1983) Lower-tail approximations for the probability of failure of 3-D composites with hexagonal geometry. *Proc. R. Soc. Lond. A*, **388**, 353–91.

Smith, R. L., and Weissman, I. (1985) Maximum likelihood estimation of the lower tail of a probability distribution. *J. R. Statist. Soc. B*, **47**, 285–98.

Smith, R. L. (1991) Weibull regression models for reliability data. *Reliability engineering and system safety*, **34**, 35–57.

Soms, A. P. (1985) Lindstrom–Madden method. *Encyclopedia of Statistical Sciences*. (N. L. Johnson and S. Kotz, eds) Wiley, New York.

Soms, A. P. (1988) The Lindstrom–Madden method for series systems with repeated components. *Comm. Statist. Theory Meth.*, **17**(6), 1845–56.

Spiegelhalter, D. J. and Smith, A. F. M. (1982) Bayes factors for linear and log-linear models with vague prior information. *J. R. Statist. Soc. B*, **44**, 377–87.

Springer, M. D. and Thompson, W. E. (1966) Bayesian confidence limits for the product of *N* binomial parameters. *Biometrika*, **53**, 611–3.

Stacy, E. W. (1962) A generalization of the gamma distribution. *Ann. Math. Statist.*, **33**, 1187–92.

Stigler, S. M. (1973) Linear functions of order statistics with smooth weight functions. *Ann. Statist.*, **2**, 676–93.

Sudakov, R. S. (1974), On the question of interval estimation of the index of reliability of a sequential system. *Engineering Cybernetics*, **12**, 55–63.

Suh, M. W., Bhattacharyya, B. B. and Grandage, A. (1970) On the distribution and moments of the strength of a bundles of filaments. *J. Appl. Prob.*, **7**, 712–20.

Sweeting, T. J. (1984) Approximate inference in location-scale regression models. *J. Amer. Statist. Assoc.*, **79**, 847–52.

Sweeting, T. J. (1987a) Approximate Bayesian analysis of censored survival data. *Biometrika*, **74**, 809–16.

Sweeting, T. J. (1987b) Discussion of the paper by D. R. Cox and N. Reid. *J. R. Statist. Soc. B*, **49**, 20–1.

Sweeting, T. J. (1988) Approximate posterior distributions in censored regression models. *Bayesian Statistics 3*, Proceedings of the 3rd Valencia International Meeting, 791–9, Oxford Univ. Press.

Sweeting, T. J. (1989) Discussion of the paper by J. I. Ansell and M. J. Philips. *Appl. Statist.*, **38**, 234–5.

Takahasi, K. (1965) Note on the multivariate Burr's distribution. *Ann. Inst. Statist. Math.*, **17**, 257–60.

Thompson, W. A. Jr. (1988) *Point Process Models with Applications to Safety and Reliability*. Chapman and Hall, London.

Tierney, L.-J. (1981) The asymptotic time of failure of a bundle of fibres with random slacks. *Adv. Appl. Prob.*, **13**, 548–66.

Tierney, L. -J. (1982) Asymptotic time of fatigue failure of a bundle of fibres under local load sharing. *Adv. Appl. Prob.*, **14**, 95–121.

Tierney, J. and Kadane, J. B. (1986) Accurate approximations for posterior moments and marginal densities, *J. Amer. Statist. Assoc.*, **81**, 82–6.

Titterington, D. M., Smith, A. F. M. and Makov, U. E. (1985) *Statistical Analysis of Finite Mixture Distributions*. Wiley, London.

Triner, D. A. and Phillips, M. J. (1986) The reliability of equipment fitted to a fleet of ships in *9th Advances in Reliability Technology Symposium*.

Tsiatis, A. (1975) A nonidentifiability aspect of the problem of competing risks. *Proc. Nat. Acad. Sci. U.S.A.*, **72**, 20–2.

Turnbull, B. W. (1974) Nonparametric estimation of a survivorship function with doubly censored data. *J. Amer. Statist. Soc.*, **69**, 169–73.

Turnbull, B. W. (1976) The empirical distribution function with arbitrarily grouped, censored and truncated data. *J. R. Statist. Soc. B*, **38**, 290–5.

Van Dyke, P. and Hedgepeth, J. M. (1969) Stress concentrations from single-filament failures in composite materials. *Text. Res. J.*, **39**, 618–26.

Wald, A. (1950) *Statistical Decision Functions*. Wiley, New York.

Watson, A S. and Smith, R. L. (1985) An examination of statistical theories for fibrous materials in the light of experimental data. *J. Mater. Sci.*, **20**, 3260–70.

Weibull, W. (1939) A statistical theory of the strength of material. *Ingeniors Vetenscaps Akademiens Handligar*, Stockholm No. 151.

Weibull, W. (1951) A statistical distribution function of wide applicability. *J. Appl. Mech.*, **18**, 293–7.

Whitehead, J. (1980) Fitting Cox's regression model to survival data using GLIM. *Appl. Statist.*, **29**, 268–75.

Whitmore, G. A. (1983) A regression method for censored inverse-Gaussian data. *Canad. J. Statist.*, **11**, 305–15.

Wightman, D. W. and Bendell, A. (1985) The practical application of proportional hazards modelling in *Proc. 5th Natl Reliability Conference, Birmingham*. Warrington, National Centre of Systems Reliability.

Williams, J. S. and Lagakos, S. W. (1977) Models for censored survival analysis: constant-sum and variable-sum methods. *Biometrika*, **64**, 215–24.

Winterbottom, A. (1980) Asymptotic expansions to improve large sample confidence intervals for system reliability. *Biometrika*, **67**, 351–7.

Winterbottom, A. (1984) The interval estimation of system reliability from component test data. *Operations Research*, **32**, 628–40.

Wolstenholme, L. C. (1989) Probability models for the failure of fibres and fibrous composite materials. Ph.D. Thesis, University of Surrey.

Wolstenholme, L. C. and Smith, R. L. (1989) Statistical inference about stress concentrations in fibre-matrix composites. *J. Mater. Sci.*, **24**, 1559–69.

Wright, B. D., Green, P. J. and Braiden, P. (1982) Quantitative analysis of delayed fracture observed in stress tests on brittle material. *J. Mater. Sci.*, **17**, 3227–34.

Zweben, C. and Rosen, B. W. (1970) A statistical theory of material strength with application to composite materials. *J. Mech. Phys. Solids*, **18**, 189–206.

Author index

Subject index